Earth Science
Discovering the secrets of the earth

GEOLOGICAL TIME

Grolier Educational

Grolier Educational

First published in the United States in 2000 by Grolier Educational, Sherman Turnpike, Danbury, CT 06816

Author
Brian Knapp, BSc, PhD

Art Director
Duncan McCrae, BSc

Editors
Mary Sanders, BSc and Gillian Gatehouse

Illustrations
David Woodroffe and Julian Baker

Designed and produced by
EARTHSCAPE EDITIONS

Reproduced in Malaysia by
Global Colour

Printed in Hong Kong by
Wing King Tong Company Ltd

Library of Congress Cataloging-in-Publication Data
Earth science
        p. cm.
      Includes index.
        Contents: v. 1. Minerals — v. 2. Rocks — v. 3. Fossils — v. 4. Earthquakes and volcanoes — v. 5. Plate tectonics — v. 6. Landforms — v. 7. Geological time — v. 8. The earth's resources.
        ISBN 0-7172-7499-3 (set: alk. paper) — ISBN 0-7172-9493-5 (v. 1: alk. paper) — ISBN 0-7172-9494-3 (v. 2: alk. paper) — ISBN 0-7172-9495-1 (v. 3: alk. paper) — ISBN 0-7172-9496-X (v. 4: alk. paper) — ISBN 0-7172-9497-8 (v. 5: alk. paper) — ISBN 0-7172-9498-6 (v. 6: alk. paper) — ISBN 0-7172-9499-4 (v. 7: alk. paper) — ISBN 0-7172-9502-8 (v. 8: alk. paper)
        1. Earth science—Juvenile literature. [1. Earth science.] I. Grolier Educational Corporation.
QE29.E27 2000
550 — dc21   99-086995
                                                    CIP

Acknowledgments
The publishers would like to thank the following for their kind help and advice: *Anne and Ron Handy* for the photograph of stromatolites, *Anna Cooper* for the sample of Precambrian Zebra Rock, and *Christopher R. Scotese* for the paleogeographic maps.

Picture credits
All photographs are from the Earthscape Editions photolibrary except the following:
(c=center t=top b=bottom l=left r=right)
Paleogeographic Maps by Christopher R. Scotese, PALEOMAP Project, University of Texas at Arlington (www.scotese.com)
22b, 32b, 34b, 36b, 40b, 42b, 46b, 48b, 50b, 55b, 57b; NASA 23, 27tr, 27b, 29, 54b.

Contents

Chapter 1: The staircase of time

This book is about geological time, a vast span of 4.6 billion years of the earth's history since its formation.

Finding out about events since the earth appeared may seem like an impossible task. But fortunately, by studying the clues hidden in rocks that can still be found at the earth's surface, it is possible to know when the oceans first came into existence, how and when the atmosphere emerged and changed, when the first living things appeared, and how the surface of the earth has altered through its long history. To achieve all of this from rocks, many ideas from earth science have to be brought together; and so in this chapter we will introduce some of the ideas of geological time by looking at a single example found in the lands around the Grand Canyon—one of the most spectacular places on earth.

The Grand Staircase

Finding out about the history of the world means building up a picture of all the rocks that have been laid down over the billions of years of the earth's history. Each layer of rock in the earth is called a STRATUM (plural strata). Studying the history of the earth through the strata is called STRATIGRAPHY.

(Below) This diagram shows the main rocks of the Grand Staircase in the southwestern United States. As you can see, erosion has stripped back layer after layer of the rocks, so that they are now easy to study.

The stepped nature of the landscape gave rise to the term Grand Staircase, which covers all of the rocks from the Grand Canyon northward across the Colorado Plateau.

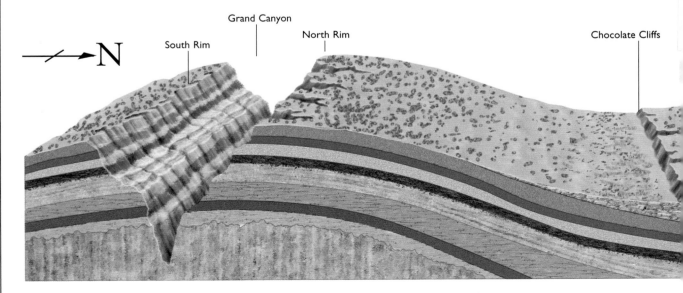

Grand Canyon

South Rim

North Rim

Chocolate Cliffs

N

If all of the rocks that ever existed were piled on top of one another, they would rise many hundreds of thousands of meters. No place exists where the whole of the history of the earth can be viewed together because the earth's rocks are continually being made and then eroded. But just occasionally, rocks remain undisturbed for very long periods of time, and then

(Below) A view over the Grand Canyon looking from the South Rim to the North Rim.

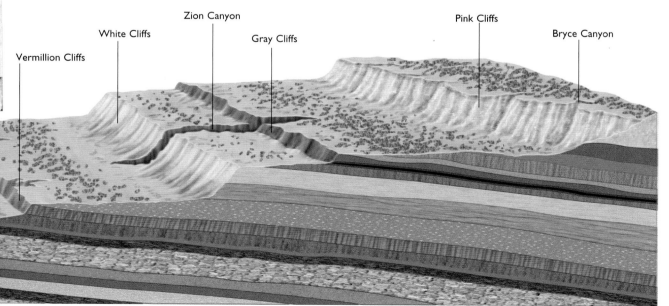

Vermillion Cliffs
White Cliffs
Zion Canyon
Gray Cliffs
Pink Cliffs
Bryce Canyon

they provide the vital clues that help us work out the history of geological time. Geologists know one of these important places as the Grand Staircase.

The Grand Staircase includes the rocks of the high plateau lands of the southwestern part of the United States. Here you can see layer after layer of rock one on top of the other. The Grand Staircase came about because for over a billion years more rocks were laid down on average than were worn away. No great earth movements folded and twisted the rocks, so that, apart from gentle swells, they are still almost level and unaltered from the time when they were laid down.

(Below) The rocks of the Grand Canyon. The names on the left refer to geological periods (see page 21), and those on the right refer to the groups of strata within each period, known as formations (e.g., Toroweap Formation). If the formation is represented by a distinctive rock texture (e.g., limestone, shale), then this name may be used instead of the word formation (e.g., Kaibab Limestone). (Diagram not to scale.)

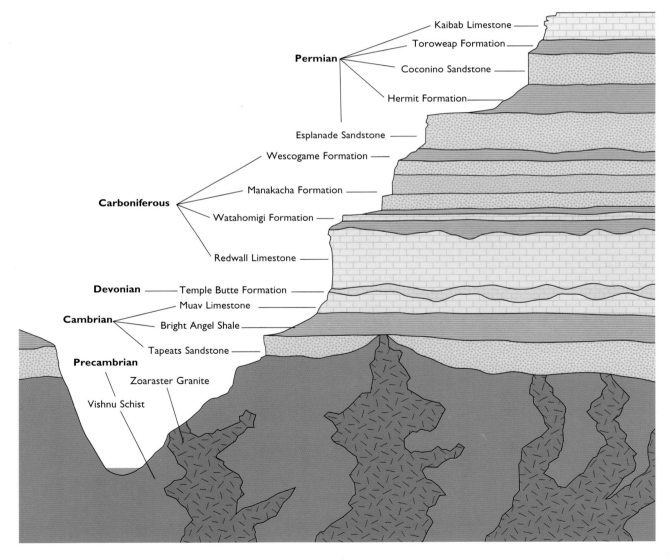

Permian
- Kaibab Limestone
- Toroweap Formation
- Coconino Sandstone
- Hermit Formation

Esplanade Sandstone

Carboniferous
- Wescogame Formation
- Manakacha Formation
- Watahomigi Formation
- Redwall Limestone

Devonian — Temple Butte Formation

Cambrian
- Muav Limestone
- Bright Angel Shale
- Tapeats Sandstone

Precambrian
- Zoaraster Granite
- Vishnu Schist

The oldest rocks

The Grand Staircase was produced by the Colorado River, which has cut down through all of the rocks of the high plateau for the last ten million years. The effect has been a peeling back of layer after layer of earth history, leaving them as a staircase of spectacular cliffs.

There are many kinds of rock in the staircase. The oldest are at the bottom, just where the Colorado River cuts into the floor of the Grand Canyon. They are the only rocks that have been twisted, folded, and changed by heat and pressure. This tells us they were once part of a mountain chain. When geologists first visited this area, they gave this group of similar rocks (a rock FORMATION) a local name—the Vishnu

(Below) The main formations of the Grand Canyon and how they span the geological time scale.

Kaibab Limestone

Toroweap Formation

Coconino Sandstone

Hermit Formation

Esplanade Sandstone

Wescogame Formation

Manakacha Formation

Watahomigi Formation

Redwall Limestone

Temple Butte Formation

Muav Limestone

Bright Angel Shale

Tapeats Sandstone

Vishnu Schist

Formation. The main rocks of the formation are **METAMORPHIC ROCKS** called **SCHISTS**.

Schists are not the only rocks in the bottom of the canyon. Thick veins of **GRANITE** cut through the schists, something that possibly happened at the height of **MOUNTAIN BUILDING**. The granite has been dated as 1.3 billion years old, so we know that mountains were forming at this time. It takes time for the layers of rock to be laid down that are eventually squashed up into a mountain range. So, the rocks at the bottom of the Grand Canyon are probably about 2 billion years old.

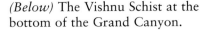
(Below) The Vishnu Schist at the bottom of the Grand Canyon.

The top of the Vishnu Schist is not a flat surface, but rises and falls. Before the rocks above it were laid down, this rising and falling was hills and valleys. Here, then, we can see a landscape some 600 million years old in which the ancient mountains were worn down to hills and valleys. Because there are other rocks on top of this landscape, the land must have sunk below sea level, allowing the ocean to flood in and cover the ancient land with layer upon layer of sediments. These sediments were eventually compressed into **SANDSTONE**, **SHALE**, and **LIMESTONE**. Notice that there are no rocks in the canyon dating between when the mountains were formed and when the ancient landscape was finally buried. This is a gap in the record of the rocks. A break of this kind is called an **UNCONFORMITY**, and in this case it represents a period of about 700 million years.

The bands of sandstone, limestone, and shale began to be laid down more than 600 million years ago. However, even though the beds lie one on top of another in level sheets, there are three periods when no rocks were laid down. This means that

The principle of cross-cutting

If a rock cuts across other rocks, it must be younger than all the other rocks it cuts across. This is the case with the granite shown in this diagram—it cuts across the schist.

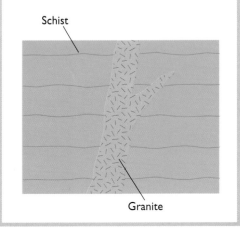

Schist

Granite

although the region was mainly below sea level, from time to time it was uplifted, and the uppermost rocks eroded away before the land sank again and new sediments were deposited once more.

The lower beds, including the formation called the Bright Angel Shales, contain evidence of some of the earliest forms of INVERTEBRATE LIFE (animals with external skeletons). The seas then contained creatures called trilobites. The number of species of life found in these sediments is rather small, showing that at this time life had not diversified very much.

Close to the top of the canyon is the Coconino Sandstone. At this point the rocks tell of a region of sand dunes (a sand sea). Even though it was a desert, FOSSILS in this formation show a much greater variety of life than in the lower rocks. They include 27

(Below) The middle part of the canyon sides is made up of layer after layer of shale, sandstone, and limestone. Sandstone and limestone are harder than shale and so stand out as vertical cliffs, while the softer shale makes debris-covered slopes.

species of four-footed animals. Thus, by moving up the rocks of the Grand Staircase, we can see evidence of a change in environments and of changes in life on earth.

The uppermost rocks of the Grand Canyon are made from the very distinctive white Kaibab Limestone. They are the youngest rocks in the Grand Canyon, formed when the land (now over 2,000 meters above sea level) was under a shallow sea.

The younger rocks

The upper part of the staircase lies outside the Grand Canyon. It, too, contains bands of limestone, shale, and sandstone, just as the lowest rocks did. This tells us that the same kinds of environment existed throughout most of geological time.

Each resistant limestone or sandstone stratum is marked by a line of cliffs: Chocolate Cliffs, Vermillion Cliffs, White Cliffs, and Gray Cliffs. The last two are cut through by Zion Canyon. Above them, in turn, are the Pink Cliffs, of which one part is split by Bryce Canyon. They were formed during the time called the Mesozoic.

(Below) White Kaibab Limestone, the youngest rock formation in the Grand Canyon.

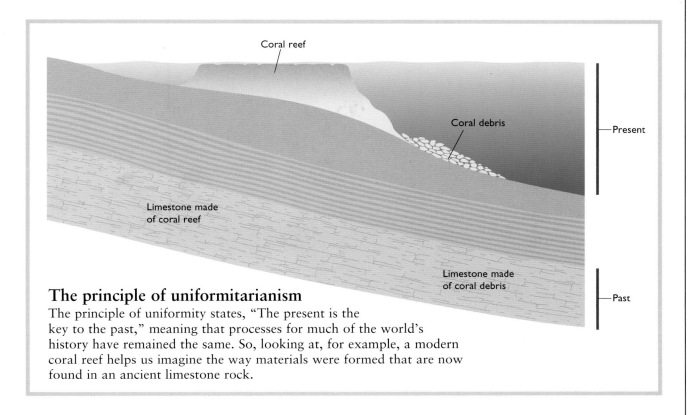

Coral reef

Coral debris

Present

Limestone made
of coral reef

Limestone made
of coral debris

Past

The principle of uniformitarianism

The principle of uniformity states, "The present is the
key to the past," meaning that processes for much of the world's
history have remained the same. So, looking at, for example, a modern
coral reef helps us imagine the way materials were formed that are now
found in an ancient limestone rock.

This is where the history in the rocks of the
Grand Staircase stops—so far, for in the future the
land may sink again, be covered by seas, and even
more rocks added.

Divisions of time

The Grand Staircase holds a record of about a third
of the entire history of the earth hidden in its rocks.
By looking at the rocks, we can tell when there were
mountains, when the land was eroded to hills and
valleys, and when it sank below the sea. By looking
at the materials in the rocks, we can see that the
environment changed, that it was sometimes a sandy
shore or a desert, sometimes a coral reef, sometimes
a muddy ocean floor. By looking at the fossils in the
rocks, we can tell that life changed. All of these are
key observations in finding out about geological time.

Because we are dealing with a very complicated
history, just as historians divide up the human
timespan to study it, so geologists find it convenient
to divide up geological time. Geologists use the word

ERA for the longest stretches of time. Smaller divisions within each era are called geological PERIODS. In the case of the Grand Staircase the eras are called (from bottom to top) the PRECAMBRIAN, the PALEOZOIC, and the MESOZOIC (see page 21). Except for the bottom of the inner canyon, all the rocks in the Grand Canyon belong to the Paleozoic Era. The time divisions of the Paleozoic range from the CAMBRIAN at the bottom to the PERMIAN at the canyon rim.

Although the Grand Staircase can be used to find out much about the history of the earth, studying just this one place cannot tell us everything about geological time. For this we need to look at rocks worldwide and then pool the knowledge. How this was done, and how the geological eras and periods were named, is described in the next chapter.

(Below) The Vermillion Cliffs. Notice that the strata can be picked out by the differences in their colors. Color changes occur as a result of differing amounts of iron and other staining substances.

(Right) The uppermost part of the Grand Staircase is being eroded by rivers to form Bryce Canyon. In the Grand Staircase the land is so spectacular that parts of it have been designated as national parks: Grand Canyon, Zion, and Bryce.

Chapter 2: How the time scale was developed

Developing a time scale for the earth has not been easy; it has taken many centuries of patient effort. Today, geologists use terms like Precambrian and Permian as a convenient shorthand for describing great chunks of time. But in the early years no one had a clear concept of geological time or names for its divisions. In this chapter we show how people came to find out about geological time and the reasons for the names that we now use.

Beginning to understand time

The starting point for the timeline began thousands of years ago, when the ancient Greeks interpreted marks in many rocks as FOSSILS—the remains of ancient creatures. They had made the first vital step to recognizing that rocks were laid down over long periods of time.

In 1669 Nicolaus Steno of Denmark examined the rocks of Tuscany in Italy and concluded that the lower rocks were older than those above. We now call this the LAW OF SUPERPOSITION, a foundation of modern geology. Steno also saw rocks tilted at an angle (DIPPING). He thought that rocks were laid down in flat layers, so that any rock layers now found dipping must have been changed since their formation. In this way he made a vital step in our understanding that rocks can be changed after they were first made.

Starting a classification

In 1756 Johann Lehmann studied rocks in southern Germany and found that some contained fossils while others did not, and that some layers below the soil were hard while others were made of loose materials. Because fossils were a sign of life on earth,

The law of superposition

The principle that, in general, rocks lower down a sequence are older than those higher up is called the principle of superposition. We can see this principle at work in rivers, in deserts, in lakes, and in oceans, with fresh, young sediment settling on top of older sediments to make new layers that will, in turn, be covered by even younger sediment.

Youngest

Oldest

he thought that rocks without fossils must have been formed before there was life on earth.

He decided that the world must contain three quite different types of rocks. The rocks without fossils were the oldest. He put them into a group called **PRIMARY** (meaning first). Rocks containing fossils that were younger than these he placed in a group called **SECONDARY**, meaning second.

Finally, he put the loose materials, such as alluvium laid down by rivers and sandy beaches, into a group called surface materials. They were the youngest materials.

In this way Lehman divided up the world's rocks into three different groups based on a combination of time and texture. It was the first attempt at a classification, and yet it is still the one on which the modern classification is based.

Giovanni Arduino then suggested that the rocks with fossils should be separated into younger, softer rocks and harder, older rocks. He called these younger rocks **TERTIARY** (meaning third).

All of the unconsolidated rocks on the surface were, at this time, still thought of as part of the Tertiary, but it was later proposed to separate them off and call them the **QUATERNARY** (fourth).

By the late 19th century it was suggested that since the oldest rocks contained fossils of the most primitive kinds of life, what had once been called the Primary Era should be renamed the **PALEOZOIC ERA** (meaning early life forms). Similar reasoning led to the words **MESOZOIC ERA** (middle life forms) and **CENOZOIC ERA** (recent life forms) to include both the Tertiary and Quaternary periods.

It is this combination of terms that is used throughout the world today.

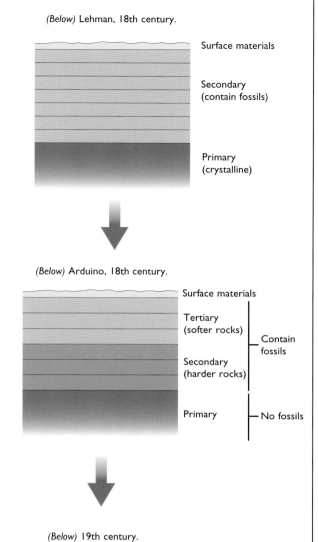

(Below) Lehman, 18th century.

- Surface materials
- Secondary (contain fossils)
- Primary (crystalline)

(Below) Arduino, 18th century.

- Surface materials
- Tertiary (softer rocks) — Contain fossils
- Secondary (harder rocks)
- Primary — No fossils

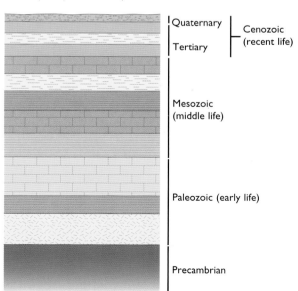

(Below) 19th century.

- Quaternary — Cenozoic (recent life)
- Tertiary
- Mesozoic (middle life)
- Paleozoic (early life)
- Precambrian

Making more use of fossils

In the 18th century James Hutton, working in Scotland, had already recognized that natural processes can produce many different kinds of sediment. For example, a river carries all kinds of sediment to the coast, but the coarser and heavier sands are deposited to make a delta, while the finer and lighter muds are often carried into the sea and deposited there as muds. Different sediments that are the same age are called FACIES.

Once it was recognized that a rock could be a sandstone in one place (a sandy facies) and elsewhere might merge into a shale (a muddy facies), people realized that using rock textures to relate rocks to one another was not the most useful method.

However, although texture might change from sandstone to shale, the same fossils could be found in both facies if the rocks were of the same age. William Smith, working in England, for example, discovered that certain fossils were always found in the same strata, and as a result he began to use the fossil contents of the rocks rather than their textures and colors, to identify rocks of similar age.

In the 19th century the ideas of evolution suggested by Charles Darwin were readily adopted

(Below) Hutton suggested that many kinds of material can be deposited at the same time but in different places.

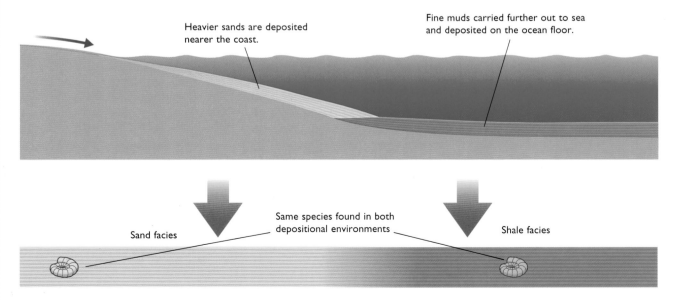

Heavier sands are deposited nearer the coast.

Fine muds carried further out to sea and deposited on the ocean floor.

Sand facies

Same species found in both depositional environments

Shale facies

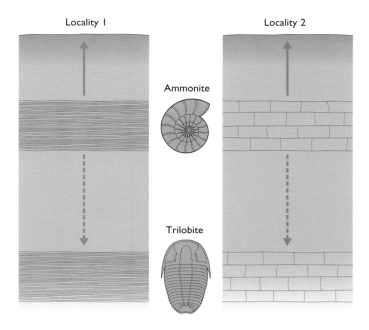

Locality I

Ammonite

Trilobite

Locality 2

(Left) Smith discovered that the same fossils could be found in many kinds of rock, and that it was better to use fossils to identify rocks of the same age rather than the colors or textures of rocks.

by the people studying rocks. As a result, the link between the evolution of species and the formation of rocks was made.

In this way it was possible to arrange fossils to make a continuous succession throughout time; and although some species became extinct, their place was taken by others. Much of this work was developed by Charles Lyell.

Naming the geological periods

As soon as attention moved to fossils, it was possible to subdivide geological time more precisely. At first, this happened in a very piecemeal way. That is why some of the first rocks to be named were the softer ones from which it was easiest to extract fossils. Much early work was done in the soft rocks of the Paris Basin in France, whose rocks belong to the Cenozoic Era.

The first rocks of the Mesozoic Era to be examined were the limestones of the Jura Mountains of France. The rocks contained similar fossils, and the geological period to which they belong is now known as the **JURASSIC**, named for the Jura Mountains.

Coal was of great economic importance, and its

(Below) A Cambrian trilobite.

fossils were eagerly examined. They, too, were widely spread throughout Europe and were soon grouped into the Carboniferous Period, named for the carbon from which coal is made.

The chalk that is widespread in France was called the Terrain Crétacé, and thus the geological period that contained chalk rocks was soon known as the **CRETACEOUS PERIOD**. Finally, periods within the Mesozoic were completed by naming the earliest period the **TRIASSIC** Period because it seemed to be divided into three.

Most of the early work on the oldest rocks with fossils was done by two geologists, Adam Sedgwick and Roderick Murchison.

Working in Wales, Murchison found three distinctively different periods containing fossils that were older than those of the Carboniferous. He called the oldest period the **CAMBRIAN**, after the Latin word for Wales, *Cambria*. Another period was called the **SILURIAN** for the Roman name for an ancient Celtic tribe, the *Silures*. The period in between them was called the **ORDOVICIAN**, using another Roman name for an ancient Celtic tribe, the *Ordovice*s.

When Murchison and Sedgwick traveled to the southern English county of Devon, they found from examining the fossils that the rocks were younger than the Silurian, but even older than the Carboniferous. These rocks had originally been placed in the bottom of the Carboniferous and called Old Red Sandstone simply because they were old rocks that were mainly red sandstones. Now a new name was proposed for them. It was the **DEVONIAN** Period, named for the county where the research took place.

Finally, Murchison went to a city called Perm in Russia to study rocks there and found that the fossils were younger than the Carboniferous. So, he had another geological period to name, and this time he used the name of the nearby city, which is why the youngest geological period of the Paleozoic Era is called the **PERMIAN** Period.

(Above) Fossil fern from the Carboniferous Period.

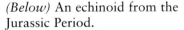

(Below) An echinoid from the Jurassic Period.

Putting dates in the time scale

The whole of the geological time scale had been worked out by using the fossil record. It was a relative time scale, meaning that the divisions were based on major changes in the fossils without any idea of when these changes happened. There were problems with this, however, especially because the fossil record makes up only 13% of the history of the earth. The 87% of earth's history that took place before the widespread preservation of living things as fossils could not be classified, and no more could be done than to group it as "Precambrian."

The first date for a sample of rock using a technique of **RADIOACTIVE DATING** was obtained by John Strutt in 1905. What he found completely revolutionized ideas about the world. He discovered that the piece of rock was 2 billion years old.

This extraordinary finding was important because it suggested that the history of the earth was far, far longer than anyone had imagined. But just as importantly, it meant that there was now proof of an enormously long period of geological time over which evolution could have taken place. By 1911 Arthur Holmes had established that rocks from the Carboniferous were 340 million years old, Devonian rock samples were 370 million years old, and so on.

As people had suspected, the dates of significant changes in the fossil record were not evenly spread over time. Nevertheless, the geological time scale was not altered because it still represented the most practical way of studying rocks. As a result, the relative time scale, which separates geological periods of uniform conditions by catastrophes when huge numbers of living things became extinct, is still the basis of the geological time scale today. Absolute dates have been especially important, though, in enabling us to understand where continents were in the past. And this is what we will look at in the next chapter.

(Above) An ammonite from the Cretaceous Period.

(Below) A gastropod from the Tertiary Period.

Chapter 3: Geological time

The geological column on page 21 is the cornerstone of geology. In this chapter you will find more detailed information about each of the eras and the periods that it contains.

How the geological time scale is organized

The geological time scale is normally drawn as a column of rocks (the GEOLOGICAL COLUMN), as though all the rocks could be found on top of one another in one place. The divisions of this column are based on major changes in the fossil record— for example, times when many species became extinct or rapid evolution occurred.

The fundamental division of the column is into GEOLOGICAL SYSTEMS, representing the BEDS of rock laid down during times when life evolved smoothly. The period of time during which the system was laid down is called the GEOLOGICAL PERIOD.

Geological periods represent very long spans of time (tens of millions of years), and they are often conveniently split into two (upper and lower) or three (upper, middle, and lower) EPOCHS. Layers of rock deposited in an epoch make up a SERIES.

The geological systems and geological periods use the same name (e.g., the rocks that make up the Silurian System were laid down in the Silurian Period; the rocks that make up the Permian System were deposited during the Permian Period).

Distinctive parts of systems are called GROUPS (if they contain beds of many different textures) and FORMATIONS if they are primarily of one material. Groups and formations are normally given local names (for example, Toroweap Formation; see page 7).

Groups of periods are called ERAS. Thus the Paleozoic Era contains the Permian, Carboniferous, Devonian, Silurian, Ordovician, and Cambrian geological periods.

It is not easy to define the boundaries between the many eras, periods, and epochs. Different authorities have their own views, and so the dates chosen as the boundaries by one may not exactly coincide with the dates chosen by another. This is why, if you read a number of books on earth science, you may find a variety of dates given for the same period of time.

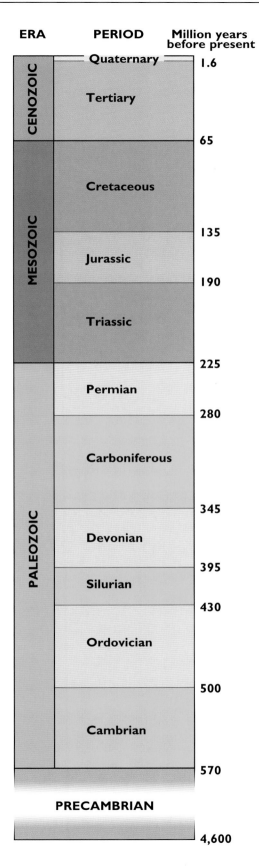

ERA	PERIOD	Million years before present
CENOZOIC	Quaternary	1.6
	Tertiary	
		65
MESOZOIC	Cretaceous	
		135
	Jurassic	
		190
	Triassic	
		225
PALEOZOIC	Permian	
		280
	Carboniferous	
		345
	Devonian	
		395
	Silurian	
		430
	Ordovician	
		500
	Cambrian	
		570
PRECAMBRIAN		
		4,600

Variations

Because the geological periods were named after the places where they were first studied, the names were not always appropriate when used in other continents. American geologists, found, for example, that the boundary between the European Upper and Lower Carboniferous Epochs did not entirely match the one they needed in North America. As a result, many American scientists use the name Pennsylvanian (named by Alexander Windchill) instead of Upper Carboniferous and the name Mississippian (named by Henry Williams) instead of the Lower Carboniferous.

The Precambrian Era

The Precambrian is the name of the largest division of geological time. The Precambrian stretches over 87% of the earth's history and contains the rocks formed before there were widespread and easily identified traces of fossil animals. Fossils became common in the Cambrian Period, starting about 570 million years ago, and so everything from the formation of the earth to the Cambrian Period is called the Precambrian.

The oldest evidence of rock material that has been found so far dates back to 4.3 billion years before the present, that is, just 400 million years after the beginning of the earth's formation.

The earliest part of earth history was its most dramatic. Before 4.6 billion years ago the earth only existed as solid grains moving in space. Gradually, gravitational and electrical attraction made these particles clump together at an increasing rate until they became the chunks of rock that astronomers call **PLANETISMALS**. Many of them we would now

570 million to 1 billion years ago	Sponges and primitive arthropods
2 billion to 3 billion years ago	First stromatolites
3 billion to 3.4 billion years ago	First primitive cells
4.6 billion years ago	Earth formed

> **Note:** The terms "Upper" and "Lower" refer to the relative location of rock strata, while "Early" and "Late" refer to time.

(Below) The world as it may have looked 650 million years ago.

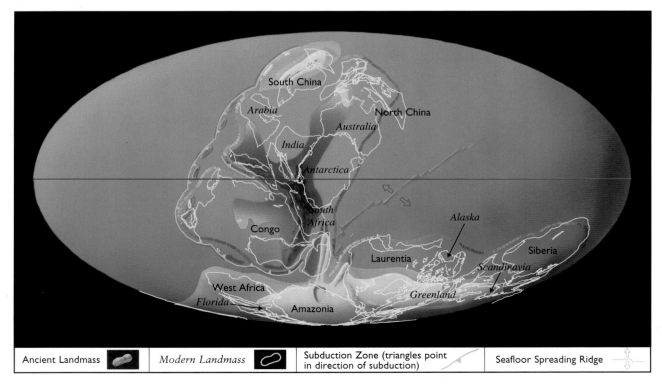

South China
Arabia
North China
Australia
India
Antarctica
South Africa
Congo
Alaska
Siberia
Laurentia
Scandinavia
West Africa
Greenland
Florida
Amazonia

Ancient Landmass	*Modern Landmass*	Subduction Zone (triangles point in direction of subduction)	Seafloor Spreading Ridge

call METEORITES. One of these planetismals grew larger than its neighbors and gradually swept up more and more material as it orbited the sun. Quite quickly the planet grew into the earth.

As the earth began to grow, it heated up, in part from the energy of many particles coming together and in part from the radioactivity given out by the particles that it contained. The earth eventually became so hot that it was entirely molten. At this time the heavier metal elements sank toward the center of the earth, making the core, while the lighter elements like hydrogen, oxygen, and other nonmetals rose to the surface.

At the same time, elements combined into compounds on the surface and cooled there enough to become solids. They formed the first CRUST. In

(Below) This folded range of mountains (seen as a darker green-brown area in the center of the image) trends generally east-west and is bounded north and south by two river valleys. The Hamersley Range has peaks that reach 1,200 meters above sea level.

The Hamersley area is a major source of iron formations, and the area north of the Hamersley contain some of the oldest rocks in the world, dating back 3.5 billion years. The lighter-colored areas north of the Hamersley, known as the Pilbara District, contains huge granite BATHOLITHS.

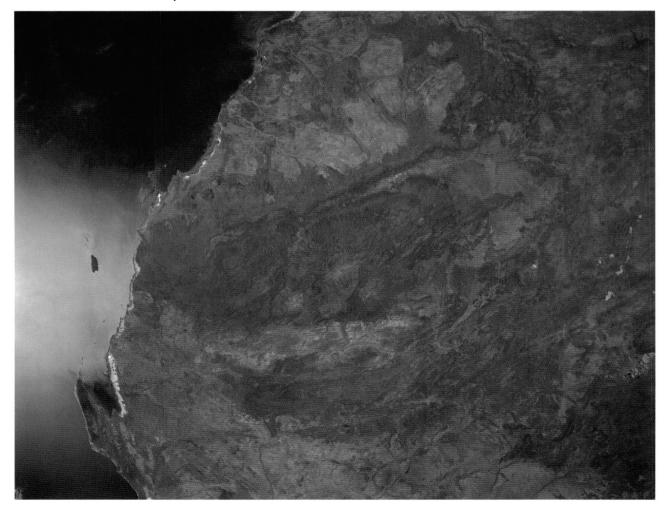

this way the earth developed its three main layers of CORE, MANTLE, and crust.

As the liquid earth lost heat to space, it formed CONVECTION CURRENTS that churned the thin crust. It probably took quite a long time for a crust to form that was thick enough to resist being pulled back into the earth by convection. The oldest minerals we find may date from rocks formed at this time.

Once the surface was sealed by a rigid crust, lava could only reach the surface through FISSURES or volcanic VENTS, just as happens at the edges of TECTONIC PLATES today.

Another important feature of the early Precambrian was the number of impacts from meteorites.

The oldest minerals still existing in the world are crystals of zircon found at Mount Narryer in Western Australia. They date to 4.3 billion year ago. The zircons occur in sedimentary rocks called CONGLOMERATES, so they must have come from yet older rocks of which there is now no trace.

The oldest intact rocks come from the Great Slave Lake in Canada and date to 3.9 billion years. Rocks dated at 3.6 to 3.7 billion years can be found in Minnesota, in western Greenland, and in southern Africa.

Formation of atmosphere and oceans

The oceans formed from the vapors pouring from volcanoes. Water vapor condensed to form the early oceans as soon as the earth's surface was cool enough. This took place within a few hundred million years of the earth's history, for without water the algae that helped change the earth's atmosphere could not have survived.

Hydrogen and helium, the lightest elements, may have risen up through the early earth while it was still molten and temporarily made the first atmosphere; but they were too light to be trapped by the earth's gravity for long, and they escaped

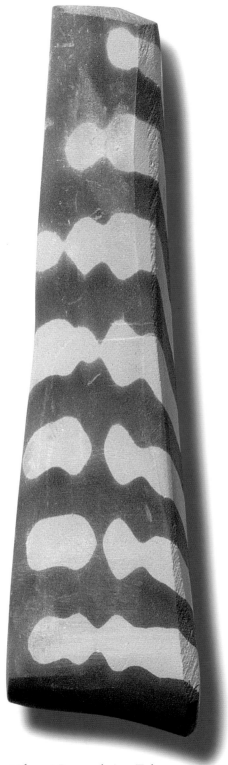

(*Above*) Precambrian Zebra Rock from Australia, a shale with bands more and less rich in iron oxide.

into space. The early atmosphere was made mostly of methane, water vapor, and carbon dioxide, and contained little oxygen. The atmosphere we breathe today is much younger and is the result of volcanic activity.

Both the atmosphere and oceans began to form in Precambrian times. The oxygen we now have in our atmosphere was possibly provided by solar radiation breaking up water vapor into hydrogen and oxygen gases. Then, over 3.5 billion years ago the first primitive plants

(Above) Precambrian metamorphic rocks shot through with quartzites make the walls of the Black Canyon of the Gunnison, Colorado.

Precambrian rocks

The Precambrian time is divided into two quite distinct stages, now called the **ARCHEAN** and the **PROTEROZOIC**, with a boundary at about 2.5 billion years ago. The earlier of these periods is the Archean, a time when there was no thick, stable crust to the earth, and continents were readily destroyed. The small fragments of crust that survive from these times are mainly volcanic or metamorphic rocks. During the Proterozoic the crust became thick enough to remain stable, and as plates crashed together, mountain belts were formed.

These early stages of the formation of the earth, and the enormous amount of volcanic activity that went with it, produced the majority of the mineral deposits that we use today. More than half of all the gold ever mined comes from the 2.4-billion-year-old Witwatersrand mines in South Africa.

Precambrian rocks are still found in the cores of the major continents. These areas are known as **CRATONS** or **CONTINENTAL SHIELDS**, e.g., the Canadian Shield of North America, the Baltic Shield of Western Europe, the Yilgarn and Pilbara shields of Western Australia. They have mainly been worn to low-lying ground during the enormous period through which they have been eroded. The other place where Precambrian rocks are found is in the cores of some mountain chains (for example, in the metamorphic **GNEISS** rocks of the islands of northwestern Scotland and the Adirondack

Mountains in the northeastern United States).

During Proterozoic times most of the rocks being laid down were similar to those we find forming today, including limestones, sandstones, and shales. Extensive volcanic activity was also produced from many **MAGMA CHAMBERS**. The remains of these magma chambers are found as granite rocks. One of the biggest **DIKES** from this age is the Great Dike of Zimbabwe. It is 2.5 billion years old, 480 kilometers long, and 8 kilometers thick. It contains enormous deposits of chromium and platinum ores. Another huge dike occurs in Montana. The Stillwater Dike is 2.7 billion years old, 48 kilometers long and 6 kilometers thick; it also contains platinum ore. "Swarms" of small parallel dikes are even more common. The 1.2-billion-year-old Mackenzie Dike swarm is 3,000 kilometers long and 500 kilometers thick; it crosses the whole of northwestern Canada. It represents a time when the edge of a Precambrian continent was being pulled apart, allowing magma to flow up into the rifts that formed.

During the Precambrian Era the climate varied enormously. For much of the time it was warm, but there is also evidence of major ice ages, the first occurring 2.3 billion years ago. The largest glaciation ever to have affected the earth also occurred in the Precambrian. Between 1 billion and 600 million years ago three glaciations occurred, covering almost the entire land surface as it existed at that time.

appeared in the form of algae. They could react with carbon dioxide and water to release oxygen. Ancient organisms called stromatolites give clear evidence of the presence of plants in ancient limestones.

The process of adding oxygen to the atmosphere was a slow one, and for another billion years all the sediments formed were dark gray because there was not enough oxygen in the air to turn the iron in the rocks red.

Even longer had to elapse before a wider variety of life forms had evolved—over 2 billion years.

The early earth produced many more volcanic outpourings than today, much of it BASALT and rich in iron. As these rocks were eroded, the iron was released into the oceans. Oxygen released by algae in the water allowed oxygen and iron to combine, making the iron insoluble and precipitating it as thick beds. Iron beds have never formed in such

(Above) Stomatolites were first seen in the Precambrian and have survived ever since. They are the longest surviving living thing. These are in Western Australia.

Precambrian life

Proterozoic comes from the Greek meaning "early life." This name recognizes that life formed through much of the Precambrian, although little is known about it. The earliest life forms were simple-celled organisms such as algae. The main concentrations of these algae are as stromatolites, dome-shaped towers or mats of algae built one upon the other. Stromatolites are the longest surviving example of life on earth, forming about 3.5 billion years ago and still forming today at, for example, Shark Bay in Western Australia.

Evolution was slow in these early times, but by the late Precambrian fungi, algae, and simple organisms called protozoa existed everywhere on earth. By 700 million years ago some of the algae had developed into complex forms such as seaweeds, and simple animals such as jellyfish and worms had evolved. All of these early animals were soft bodied and have left only rare traces. That is why people originally thought that the Precambrian was devoid of life. It was only when the animals developed hard shells that they began to leave the remains that are so abundant in the Cambrian Period. (A more comprehensive discussion of the fossil record will be found in the book *Fossils* in the *Earth Science* set.)

concentrations since, so that these Precambrian iron beds are the major source of the world's iron ore.

Oxygen then began to seep out of the ocean water and build up in the air. By the time of the land plants (about 400 million years ago) the atmosphere already had about the same oxygen content as it has today.

(Right) The Great Dike of Zimbabwe contains enormous deposits of valuable ores.

(Left) The Adirondack Mountains of the northeastern United States are part of the Canadian Shield that formed in Precambrian times. The pink criss-cross features are granite batholiths. Glaciation has scored long valleys across them, now occupied by lakes.

The Paleozoic Era

The **PHANEROZOIC** is the name given to the huge expanse of time called an **EON** that stretches from the Precambrian to the present. The oldest era of the Phanerozoic is called the Paleozoic. It represents what, in the past, was known as the Primary Era.

The Paleozoic Era began about 570 million years ago, when animals first developed skeletons, and ended about 225 million years ago, when there was a major catastrophe that wiped more species off the face of the earth than at any other time in history. During this time a wide variety of plants and animals lived in the oceans, plants and animals colonized the land, and insects took to the air.

The Paleozoic took up over half of the Phanerozoic, that is, nearly 350 million years. The Paleozoic Era is divided into the Cambrian, Ordovician, Silurian, Devonian, Carboniferous, and Permian periods. In North America the Lower Carboniferous is often called the Mississippian Period, and the Upper Carboniferous is often called the Pennsylvanian Period.

Summary

Paleozoic rocks are found on all continents, mainly as sedimentary rocks. In the Lower Paleozoic, fossils such as trilobites, graptolites, and ammonites became widespread and can be used as **INDEX FOSSILS** to match rocks over wide areas.

At the start of the Paleozoic the continents were widely scattered across the earth (although many were close to the equator, and none were at the poles). During the Paleozoic they slowly moved together until they formed one supercontinent.

(Below) Many Paleozoic mountains have been worn down so that their granite cores are now exposed. Around these Hercynian-aged granites are many veins rich with minerals. This granite coast of northern Cornwall, England, was mined for copper in the 19th century. The mines had to be located precariously on the edge of the granite cliffs in order to reach the copper veins that stretched under the sea.

(Above) Exposed folds of the Appalachian Mountains, dating back to the time of Caledonian Mountain Building.

In Cambrian times there were six major continents: Baltica (northern continental Europe, including western Russia), Siberia, China (southeastern Asia), Kazakhstania (central Asia), Laurentia (North America and Greenland), and Gondwana (Antarctica, Africa, South America, India, Australia). The continent that showed least signs of moving was Laurentia. It remained astride the equator throughout the Paleozoic. On the other hand, the continent known as Gondwana moved toward the South Pole and experienced heavy glaciation.

Siberia moved slowly from the equator into northern areas near the Arctic. Baltica made a very long journey from the southern hemisphere, through the equator, to the northern hemisphere, eventually colliding with Laurentia by the Devonian Period. The other continents then collided with Laurentia to make the supercontinent called Pangaea by the end of the Paleozoic.

All of these collisions created mountains along the edges of the continents. The mountains formed during this time included those known as the Caledonian, Appalachian, Hercynian, Laramide, Rocky, and Ural mountain systems of North America and Europe.

During the early part of the Paleozoic a large ocean existed between Laurentia and Baltica (just as the Atlantic does today). It has been called the Iapetus Ocean. Erosion of the continents sent huge volumes of sediment into Iapetus. At the same time, the land near the edges of the continents began to sink, creating the conditions for huge volumes of sediment to build up. This long coastal region is called a GEOSYNCLINE (geo meaning "of the earth," and syncline meaning "downfold"). The geosynclines contained the great thicknesses of sediments that

Paleozoic life

The Paleozoic Era covers three-fifths of the fossil record. By the end of the Permian Period only flowering plants, birds, and mammals were yet to appear: all other forms of living things had evolved at this early stage.

Paleozoic life is much more evident in the rocks than life of Precambrian times because many species developed hard parts (called EXOSKELETONS). In the Cambrian Period trilobites (which resemble king crabs) and brachiopods (which are shellfish) were common.

By the Ordovician Period brachiopods, corals, and bivalves are recorded. Graptolites and fish (some of the first vertebrates) also made an appearance at this time.

By the Silurian Period many more species of trilobites, corals, and brachiopods existed. At this time, too, the first land plants occurred, along with the first land animals (which looked like centipedes).

By the Devonian and Carboniferous periods land plants had developed many new species. The ferns and scale-trees were important in forming coal deposits. New invertebrates like crinoids ("sea lilies") became common in the seas. Insects were also important land animals. The earliest known reptiles appeared.

The Permian Period was a complete contrast to the warm, tropical, and wet times that had existed before. Many animals became extinct, for example, trilobites and many brachiopods. Many of the old scale-trees and ferns disappeared, and coniferous trees took their place. Reptiles saw all of this change through and started to become the dominant animal life form on earth. But change was also on its way, for the first signs of the evolution from reptile to mammal were also appearing.

would later be crushed into the Caledonian mountains of Silurian times and the Hercynian/Appalachian mountains of Devonian and Carboniferous times.

(Below) The Fraser River in Canada cuts through Paleozoic and Precambrian rocks in the heart of the Rocky Mountains. The mountains have been uplifted several times since.

Paleozoic rocks

The edges of the continents were very active throughout the Paleozoic, while the centers of the continents were stable, moving only slightly up and down. When they sank, shallow seas flooded over them, and sediments were laid down uniformly over wide areas. Some of the most common rocks were limestones (from shallow seas) and sandstones (from deserts).

During the late Carboniferous Period large areas of the continents fluctuated close to sea level. When the land rose, tropical forest swamps grew; but as the land sank, the forests were flooded and covered with muds and

sands, only for the land to rise again at some point later on. These alternations of conditions create very distinctive patterns of rock strata called CYCLOTHEMS. In them coal seams were formed.

Many of the rocks formed in the Paleozoic are important today. Unlike the Precambrian, when the most important rocks were metamorphic and igneous, the most important rocks of the Paleozoic are sedimentary. For example, limestone is widely used for building materials (stone and cement), and coal is used as a fuel.

The Cambrian Period

The Cambrian Period is a time when the existing land was rapidly eroded, and great thicknesses of sediment built up in the oceans. No major mountains were formed, but the sediments laid down during the Cambrian Period made a large contribution to the rocks of later mountains.

At the start of the Cambrian Period much of the land was quite high due to earlier periods of continent building. As rain fell on the continents, enormous volumes of material were eroded, which were carried by rivers to the oceans. On the eastern side of the ocean that separated Baltica (Europe) from Laurentia (North America) a wide part of the ocean floor began to subside. This was a geosyncline. You can imagine it as rather like the eastern side of Asia today, with a string of volcanic islands offshore, then a shallow sea between the islands and the continent.

Sediments settled in deep water on the oceanward side of the islands and in shallow water on the

(Below) The world as it may have looked in the Late Cambrian.

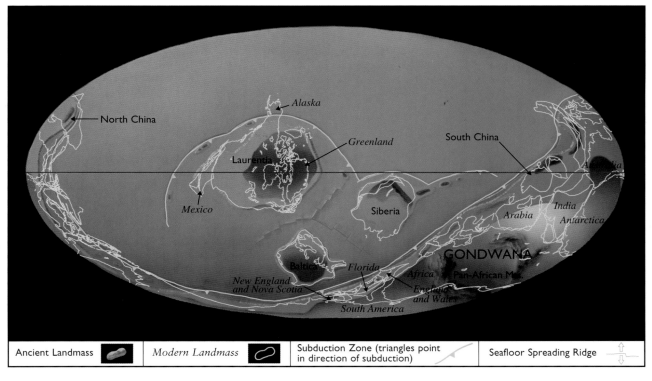

| Ancient Landmass | | *Modern Landmass* | | Subduction Zone (triangles point in direction of subduction) | Seafloor Spreading Ridge |

continental side. From time to time the volcanoes erupted, and volcanic lava and ash were added to the sediments of the geosyncline. As a result, in some places the rocks of the Cambrian are thick (deep-water) shales, while in other places there are (shallow-water) sandstones and limestones; all have intermittent beds of volcanic rocks.

Cambrian rocks

Today Cambrian rocks are still exposed in places like the Rocky Mountains in Alberta and Montana. Here, ripple marks that formed on ancient sea shores are preserved. One of the most famous locations is the Burgess Shale, which contains fossils even of soft-bodied creatures such as jellyfish. Cambrian-age sediments are also found along the borders of England and Wales (the site for which the rocks are named).

Cambrian life

Life expanded and diversified rapidly in the Cambrian because the oceans still did not contain very much life, and so there was a chance for all forms to adapt to the wide variety of ocean environments. The basis of the food chain—plants—was entirely restricted to algae and bacteria. The dominant animal life was the arthropods, of which the trilobites were most common (making up three-quarters). They evolved into thousands of species. These creatures were bottom dwellers, scavenging in the dead matter that settled on the seabed. The Cambrian sea also contained brachiopods, shellfish that made up about a fifth of ocean life.

(Below) Cambrian rocks make these imposing mountains in the Waterton-Glacier International Peace Park, in Montana.

The Ordovician Period

The Ordovician Period followed the Cambrian with continued deposits of sediments in the ocean geosynclines next to the continents. The high land of the Cambrian times was worn away, and the land everywhere was low. During this time half of Laurentia (the core of modern North America) subsided below sea level and was flooded.

One of the geosynclines filling up lay off the eastern shore of Laurentia from Greenland through Canada to the present-day Appalachians as far south as Alabama. It was part of a huge geosyncline whose rocks are also found in Europe from Scandinavia through northern Britain to Ireland. Another geosyncline lay off the western coast of Laurentia from Alaska through western Canada and the United States.

The western geosyncline was not very active and simply sagged deeper and deeper, allowing more and more sediment to accumulate. The eastern geosyncline between Laurentia and Baltica (Europe) was often

(Below) The world as it may have looked in Middle Ordovician times.

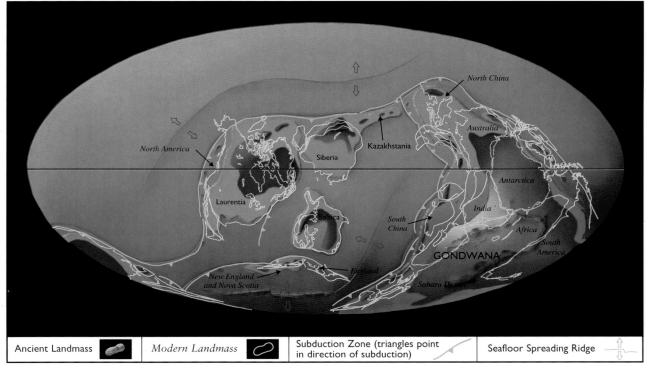

| Ancient Landmass | | *Modern Landmass* | | Subduction Zone (triangles point in direction of subduction) | | Seafloor Spreading Ridge | |

crushed, however, as the Iapetus Ocean closed and then opened again, giving rise to episodes of mountain building and volcanic activity. This is known as the **CALEDONIAN MOUNTAIN-BUILDING PERIOD** (after the Caledonian Mountains of Scotland, where the mountain system was first identified). It is also known in North America as the Taconic Mountain-Building Period, after the Taconic Mountains of New York. A great range of mountains formed that now is the core of the mountains of Scandinavia, northern Britain, Greenland, and the Appalachians.

(Below) The Highlands of Scotland date from mountain building in Ordovician times.

Ordovician life

Evolution went fast during Ordovician times. Trilobites were still numerous but beginning to decline. Brachiopods continued to evolve and adapt. Corals made more of a presence, and great reefs were formed. Bivalves became more important, some growing into giants up to 5 m long. The animals known as echinoids flourished, particularly in the form known as crinoids (sea lilies), and starfish (another echinoid) appeared. Nautiloids (which would evolve into ammonites) began to diversify into many species.

Floating in the water were colonies of small animals known as graptolites. The Ordovician has been called the Age of Graptolites because they were so widespread and common. Their free-floating life form meant that species would be found worldwide, making them very suitable for relating rocks from one continent to those from another. Their rapid evolution, producing numerous species, made them even more useful as index fossils.

Toward the end of the Ordovician many species became extinct, possibly because of widespread glaciation.

Ordovician rocks and minerals

Many Ordovician limestones are used for building materials and cement. These rocks also contain lead and zinc, worked both in North America and Europe. Many slates used for roofing also date from this time. Iron ore was formed and is now mined in Newfoundland. The decomposing remains of plants and animals were sufficiently great to produce both gas and oil fields, for example, many of the oil fields of Texas.

Mountain building produced much volcanic activity and a wide range of concentrations of metals in Gondwana, now South Africa, Antarctica, and Australia. The gold fields of Bendigo, Australia, were formed at this time.

The Silurian Period

The Silurian Period was a very short geological span in terms of absolute time, about half as long as most. Nevertheless, great thicknesses of sediments accumulated during this period, stretching across much of the continents, many of which were, in part, flooded by the sea. The results include massive beds of limestone. Because so much widespread deposition occurred in tropical seas, the Silurian Period, although brief, is marked around the world by some spectacular limestone landscapes.

During Silurian times there was a large ocean centered on the North Pole and a large continent situated at the South Pole, but stretching up to the equator. There were a number of smaller continents close to the equator.

One of the best exposures of these rocks is the cap rocks (called Lockport Dolomite) of Niagara Falls. The gorge is cut down into the softer Rochester Shales, also of Silurian age. The falls are part of the Niagara Escarpment, which is 1,000 kilometers long, reaching

(Below) The world as it may have looked in Middle Silurian times.

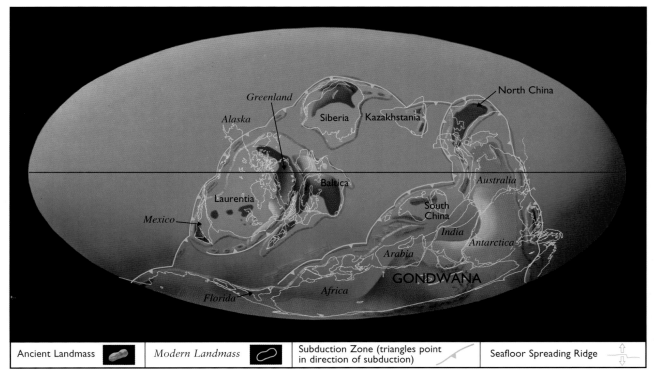

Ancient Landmass		*Modern Landmass*		Subduction Zone (triangles point in direction of subduction)	Seafloor Spreading Ridge

as far as Winsconsin, and making a cliff that overlooks the Great Lakes.

The original site for the discovery of the Silurian Period was the limestone ridge called Wenlock Edge in England. The Dingle coast of Ireland is made of Silurian rocks, while the gorges in Australia's Kalbarri National Park are also cut in Silurian rocks (Tumblagooda sandstone). There are over 200 granite intrusions in New South Wales, Australia, that were formed in Silurian times. Australia has the largest area of Silurian volcanic rocks in the world.

During Silurian times the Iapetus Ocean opened again, and a geosyncline developed down its eastern side. At the close of the Silurian the Iapetus Ocean disappeared as Laurentia and Baltica came together. The Silurian System and many older rocks from the geosynclines were then crushed together to complete the formation of the Caledonian Mountains, whose ancient rocks are found from Scandinavia to the North American Appalachians.

Silurian life

The warm, shallow seas of Silurian times gave an opportunity for species to diversify. Tabulate corals and stromatoporoids (cushionlike masses of blue-green algae) formed into vast reefs that later became great limestone beds. On the shallow sea floor brachiopods, corals, and cephalopods (squids and ammonitelike animals) thrived. Trilobites, however, were declining. Fish became more numerous and developed movable jaws. Echinoids (crinoids and starfish) were also abundant.

Most of the plants were still primitive algae, although as seaweeds they could be large. However, the first land-dwelling plants started to appear at the end of the Silurian—an important change because they provided the food that land-based animals would need if they were to survive.

On land, too, centipedes were now exploring and breathing air—possibly the first creatures to do so.

Silurian rocks and minerals

Thick beds of Silurian-age salt have been mined in the United States. The development of the Industrial Revolution can also be connected with Silurian rocks, since the creation of the first iron plant by Abraham Darby in Ironbridge, Shropshire, England, was made possible by the local presence of the Silurian-aged Wenlock Limestone (which was needed as a flux in iron-making with coal). In Birmingham, Alabama, the Silurian-aged iron ore deposits (Clinton ore) were responsible for the development of the iron and steel industry in that part of the United States.

(Left) Niagara Falls is capped by Silurian rocks.

The Devonian Period

The newly formed Caledonian Mountains of the Silurian Period (created when Laurentia and Baltica came together) played an important role in determining the character of Devonian rocks. The new continent, known as Euramerica or Laurussia, had a chain of great mountains that acted as a barrier to the flow of moisture, creating huge deserts in their rainshadow regions. Sand dunes, playa lakes, and alluvial fans are therefore characteristic of many Devonian rocks that formed close to the Caledonian Mountains. The Devonian is one of the first periods when there is any evidence of large-scale sediments building up on land. The deeply weathered red desert sandstones are so distinctive that some of the most characteristic Devonian rocks are known as Old Red Sandstone. Away from the deserts many rivers still flowed to seas and produced marine sediments as well. Thus in Devon, England, after which the period is named, the north has sandstone rocks, while the southern shore is of shales. This is a good example of facies.

Sediments flowed from this eroding continent southward and began to fill a geosyncline that formed across what is now southern Europe and Appalachia in North America.

The present-day southern continents were still joined together as Gondwana. Gondwana began to crush into Laurussia during the Devonian. The result was the crushing of the geosyncline between them and widespread mountain building that continued from the Devonian Period right through the Carboniferous Period. In time, it produced a line of

(Below) The slates that make up the Rhine Gorge in Germany were formed in Devonian times.

mountains that stretched from what is now Central Europe, through Southern Britain, to Appalachia in North America. In all of these areas granite intrusions are common (for example, Dartmoor, England). The continent called Kazakhstania (now Central Asia) also moved westward, colliding with the east of Baltica and forming the Ural Mountains.

To the west of the Appalachians the old core of Laurentia was little disturbed, and the land simply sank slightly, allowing a vast shallow sea to form around all of the edges of Laurentia except the area that is now the Midwest. Great sheets of limestone and shales formed in these conditions. In what was then the west of the Laurentian continent, a further geosyncline began to be crushed, this time resulting in the Rocky Mountains. Huge thicknesses of Devonian-aged rocks are found all the way from New Mexico to Montana and in the Sierra Nevada Mountains of California, many of them incorporating volcanic rocks and granite intrusions.

Off the coast of what is now Australia a geosyncline called the Tasman Geosyncline was forming. Much of it now makes many of the rocks along the eastern seaboard of Australia, including the Great Dividing Range.

Devonian life

During the Devonian Period there is the first widespread evidence of life on land. The first amphibians appeared, but as today, they were not entirely independent of the sea. Although they were able to live on land, they still needed to use the water as a place to lay eggs and for the young to develop.

The first large land plants also appear in the Devonian Period, giant tree ferns as much as 12 m tall, making great forests. Associated with them are the fossil remains of the first insects.

In the oceans brachiopods were common, corals formed reefs, and crinoids attached themselves to the rocky seabed. Cephalopods had up to this time had straight shells, but in the Devonian Period the shells became coiled, so that a major step in the evolution of ammonites occurred. Bony fish and the first true sharks appeared. Because fish were so abundant, the Devonian Period has been called the Age of Fishes.

Toward the end of the Devonian Period there seems to have been a widespread change in climate, which may, in turn, have been responsible for the widespread extinction of species that marks the end of Devonian times. Most trilobites, many brachiopods, and some corals all disappeared at this time. No one is exactly sure why this happened, but it could have been connected with the dust produced after the earth collided with a large meteorite.

Devonian rocks and minerals

The Old Red Sandstone is very extensive. It is used as a building stone and also as a source of sand for glass-making. The Rhine River cuts the Rhine Gorge through Devonian-aged slate rocks in Germany, and roofs of the medieval castles along its banks are clad in Devonian slates. A band of hard Devonian sandstone makes a dangerous shoal in the river that is known as the Lorelei.

The limestones of Devonian times are used both for building stone and for cement. Petroleum is also widely found in Devonian rocks and is recovered in central United States, Alberta, Canada, the North Sea, and in Central Asia.

In central Canada Devonian salt beds are mined. In southwest England and in central Europe extensive deposits of copper, tin, and zinc formed the basis of important mining industries.

The Carboniferous Period

The Carboniferous Period is famous for its vast coal swamps. They were the swamps that produced the coal from which the term "Carboniferous," meaning "carbon-bearing," is derived. Four-fifths of all the world's coal reserves come from deposits of this age.

The Carboniferous Period was not a time of uniform conditions. The Early Carboniferous produced mainly limestone rocks, while the Late Carboniferous produced swamps that turned into coal seams. The rocks of the Early Carboniferous formations make up the Lower Carboniferous (Mississippian in the United States), and the rock formations of the Late Carboniferous make up the Upper Carboniferous (also Coal Measures in Britain and Pennsylvanian in North America).

The Carboniferous Period marked some real changes in evolution, including eggs that could be laid on land (instead of being laid under water, as had previously been the case). This, in turn, paved the way for the evolution of birds and mammals.

(Below) Carboniferous age mountains of the New River Gorge in West Virginia. This southern Appalachians state also contains important coal seams.

(Below) Hard limestone is characteristic of this period. This is the "mountain limestone" of Cheddar Gorge, England.

(Below) The world as it may have looked in Early Carboniferous times.

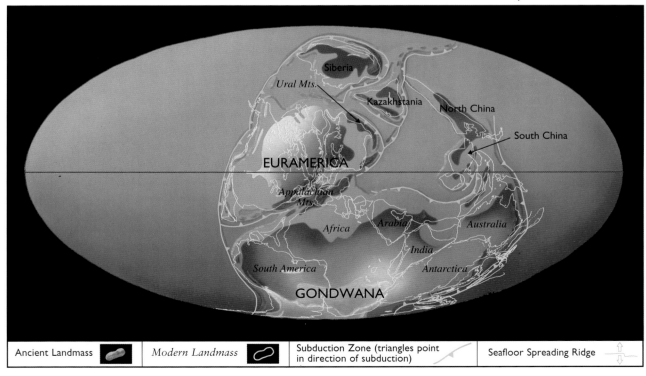

Ancient Landmass		Modern Landmass		Subduction Zone (triangles point in direction of subduction)		Seafloor Spreading Ridge	

The shallow seas of the Lower Carboniferous were an ideal place for the development of huge beds of limestone. Crinoids and algae are the most common organisms in some beds, corals and algae in others. The Upper Carboniferous environment was much more varied, with many changes of sea level caused by glaciation near the poles. Cyclothems were produced in unique conditions and have produced coal beds on many continents.

Different conditions prevailed across the great northern continent of Laurussia during the Lower Carboniferous. On the western side it was desert, with thick beds of salt and other evaporites forming, as well as a region of shallow tropical seas in which limestones formed. This region is now central and western North America. A further shallow sea covered much of what is now western Europe, producing the limestone beds that are now known as "mountain limestone," and that make such national parks as the Yorkshire Dales and the Peak District national parks of England. By contrast, on the southern continent of Gondwana there was widespread glaciation.

By the middle of the Carboniferous Period Gondwana had drifted far enough north to crush into the southern flank of Laurussia, creating the Hercynian Mountain-Building Period. Mountains formed from southern Ireland through Wales, southwestern England, and the Belgian Ardennes to Germany. In North America the same mountain building is called the Alleghenian Mountain-Building Period. The effect was to add to the southern end of the Appalachian Mountains. A further collision of Siberia into Laurussia created the Ural Mountains.

Once these continents had joined, they remained together, while the great period of fluctuating sea levels brought on by more and more extensive glaciations and meltings caused the conditions that, near the tropics, produced coal-bearing swamps and typical cyclothems.

Carboniferous life

With many species being lost at the end of the Devonian Period, a different balance of fossils is found in the Lower Carboniferous. Coral reefs are no longer widespread, and trilobites are missing. Brachiopods are less common, while bivalves and ammonoids are increasingly found. Armored fish became extinct, but other fish diversified. Crinoids are especially numerous, so that the Early Carboniferous has been called the Age of Crinoids. Most of the Lower Carboniferous limestones are made of the fragmented remains of crinoids in a limy, muddy ooze.

The first reptiles appear on land by the Late Carboniferous. On land huge tree ferns and scale-trees provided the swampy forests from which most coal is made. Insects also flourished—some much larger than anything found today—so that the Late Carboniferous has been termed the Age of Insects. Cockroaches 10 cm long and dragonflies with wingspans nearly a meter across lived in the swampy forests

Carboniferous rocks and minerals

The Carboniferous Period contains some of the most economically important rocks in the world. The Lower Carboniferous contains limestone used as building stone, as fertilizer, and as cement, while the Upper Carboniferous contains four-fifths of the world's entire coal deposits. Coal seams occur in the interior and Appalachian region of the United States, in Britain, France, Belgium, Germany, Spain, North Africa, China, Korea, and northern India.

The Permian Period

The Permian Period marks the end of the Paleozoic Era, a time when the supercontinent of Pangaea finally formed.

In many places these rocks are called red beds, the oxidized iron in the rocks that produces the color indicating that they were formed on land. Other Permian-aged rocks are thick beds of evaporites, including salt beds. They indicate basin conditions in a mountainous basin-and-range type of area. Dolomite limestones also occur at this time. These rocks make the Dolomite Mountains in Europe, a hard, almost fossil-free limestone. Much of the Permian Period has a poor fossil record.

During the Permian Period North America, Europe, Siberia, Kazakhstania, and China welded together to form a single continental block called Laurasia. North America became welded to South America, which was already part of Gondwana, so that the result was a huge supercontinent, later named Pangaea, with a split across the middle occupied by a

(Below) The world as it may have looked in the Late Permian.

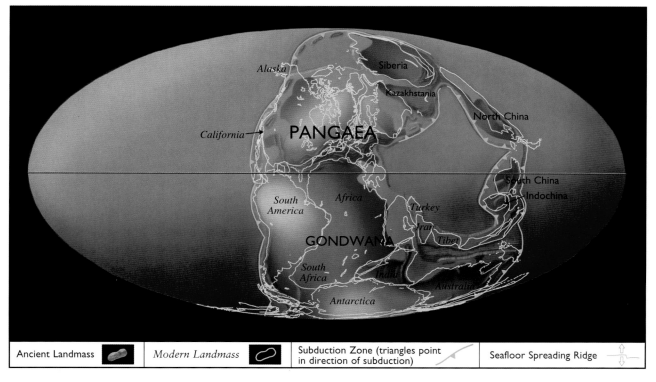

| Ancient Landmass | | *Modern Landmass* | | Subduction Zone (triangles point in direction of subduction) | Seafloor Spreading Ridge |

long ocean that has been named Tethys. Tethys lay along the equator.

The most active areas were at the edges of Pangaea. Here subduction occurred, leading to volcanic activity such as we see in South America today, only at this time it occurred on what is now the western edge of the Americas, in Antarctica, Australia, New Zealand, and along the eastern seaboard of what is now Asia.

(Below) Natural bridges, at the Natural Bridges National Monument, Utah, are formed in Cedar Mesa sandstone of Permian age.

Permian life

The Permian Period saw fewer and fewer forms of life and eventually the extinction of almost one-half of the families of the Paleozoic Era. This fact, involving total loss of many major groups, is the reason for recognizing separate Paleozoic and Mesozoic eras. The effect was most pronounced in the oceans, with less impact on land animals and plants.

Permian seas still had rugose corals as the major reef builders, sponges, and brachiopods. But at the end of the period the rugose corals were no longer dominant, while most brachiopods became extinct. This period also saw the end of the crinoids as important marine creatures.

Insects continued to evolve, while reptiles became the dominant form of life on land. The ancient seed forest and scale-trees were replaced by conifers, while the first ancestors of the flowering plants appeared.

The presence of one huge continent and one ocean led to fewer niches for many animals and plants to evolve, and the drier and cooler conditions of the time were probably responsible for the widespread extinctions that occurred and which mark the end of the Paleozoic Era.

By the end of the Permian Period conifers dominated other older plant forms.

Permian rocks and minerals

Permian rocks are especially important as sources of salt and other evaporite minerals. The salt beds of Cheshire, England, for example, were the reason for the growth of one of the world's biggest concentrations of chemical factories on nearby Merseyside. Evaporites have been similarly important in Texas. Permian deposits of petroleum are also of vital significance. The biggest Permian oil and gas fields are in West Texas (the Midland Basin), New Mexico, Oklahoma, and next to the Ural Mountains in Russia. Coal of Permian age also exists in peninsular India, Australia, China, and on the Korean peninsula.

The Mesozoic Era

The word Mesozoic means "middle life," and it is used for the era between the Paleozoic and the Cenozoic. It lasted between the Triassic Period (225–190 million years ago), through the Jurassic Period (190–135 million years ago), to the Cretaceous Period (135–65 million years ago).

Summary

During the Mesozoic Era the supercontinent of Pangaea broke up, and the positions of the continents became much more similar to those that we see today.

An early break was marked along the line that divided North America from the rest of Laurasia. Great outpourings of basaltic lava occurred along the spreading boundary. Dry conditions prevailed in the early Mesozoic, so that many of the rocks that formed on land were red. Large areas of salt deposits are also found at this early time in the Mesozoic Era.

During Jurassic times North America drifted away from what was no longer Laurasia but just Eurasia. On the "leading" edges of the westward-drifting continent subduction occurred as well as much volcanic activity, including the placement of huge granite batholiths. At the same time, the great Tethys Ocean that had split Laurasia and Gondwana began to close again as South Africa, India, and Australia began to split off from Gondwana and move north. As this happened, massive basaltic lavas began to spread over the land near the spreading boundaries.

Wherever the spreading boundaries occurred, geosynclines began to form as well as on the leading edges of the moving continents. Geosynclines formed

(Above) Layer after layer of sedimentary rocks mark the Jurassic shallow seas. Where they come to the coast they often form steep, stepped profiles, as here at Whitby, England. This is also a famous collecting site for ammonites of Jurassic age.

(Below) The granidiorite of Half Dome, Yosemite National Park, California, was put in place in Mesozoic times.

around the Pacific Ocean and also along the eastern seaboard of North America as well as on the boundaries of the Tethys Sea.

During the Jurassic Period the splitting of the continents gave more opportunity for basalt to be formed on the ocean floors. As a result of this extra amount of crust in the ocean, sea levels were forced upward, flooding many continental areas under shallow seas.

In the Late Jurassic the westward movement of North America caused a major period of mountain building along the western part of the continent. This is called the Nevadan Mountain-Building Period. It included the emplacement of huge granite masses in what are now the Sierra Nevada Mountains along the California border.

The Cretaceous Period saw the Atlantic widen further. Antarctica, Australia, and India all moved apart, the rifting being marked by more basaltic lava eruptions. The Tethys Ocean narrowed further, changing into the nearly closed Mediterranean Sea.

Mountain building began, which would later turn the Tethys Geosyncline into the Alpine and Himalayan mountain systems. As the continents subsided, there was an extraordinary flooding of the land, so that the total land above sea level was drastically reduced. Great beds of chalk and other sediments were laid down. Then, at the end of the Cretaceous Period the continents began to rise again.

Mesozoic life

During the cool, dry Triassic times the number of species had been drastically cut back, especially among invertebrates. But the change in climate also gave opportunities to other species such as conifers. Another major extinction took place at the end of the Triassic, wiping out a third of all species.

Then, when warmer times came with the Jurassic, many new species developed, including the flowering plants. In the animal kingdom the same pattern occurred, with amphibians becoming largely extinct, while reptiles increased in numbers and in species. One important group, the thecodonts, evolved into ichthyosaurs, dinosaurs, crocodiles, pterosaurs, and birds. This was the time when reptiles in particular grew to enormous sizes and occupied virtually all habitats, completely dominating life on earth.

Cretaceous conditions were, if anything, more mild, and widespread coals formed. But by the end of the period plesiosaurs, ichthyosaurs, pterosaurs, and mesosaurs had all become extinct, allowing the mammals to begin to diversify and increase in numbers and importance. For example, *Archaeopteryx*, the first bird, appeared in the Late Jurassic. By the Tertiary true birds had evolved.

Insects recovered much more slowly and hardly appear in the fossil record until the Mid-Cretaceous. They seem to evolve with the rise of flowering plants, and by the Tertiary Period they are widespread and abundant. New forms of corals appeared in the Mesozoic Era to replace the rugose corals that had almost become extinct by the end of the Paleozoic Era. Brachiopods continued to decline, but bivalves became more abundant. The Mesozoic Era was also the time when ammonites became very abundant and diverse, their free-swimming forms making them ideal for use in zoning the whole Mesozoic Era. They were the most important invertebrates of the time.

Toward the end of the Mesozoic Era many species became extinct. This was not a sudden process, but a gradual one taking millions of years to complete. Not all species became extinct at the same time, suggesting that there were several separate reasons for extinction.

The dinosaurs and the ammonites were the most conspicuous casualties.

The Triassic Period

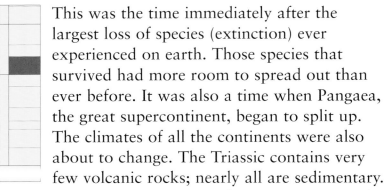

This was the time immediately after the largest loss of species (extinction) ever experienced on earth. Those species that survived had more room to spread out than ever before. It was also a time when Pangaea, the great supercontinent, began to split up. The climates of all the continents were also about to change. The Triassic contains very few volcanic rocks; nearly all are sedimentary.

The great supercontinent of Pangaea was mostly dry land. There were no inland seas, and little continental shelf was flooded. As a result, the Early Triassic contains few rocks formed under the sea. On land warm, dry climate prevailed, and the rocks formed are widespread sandstones, silts, and shales, known as red beds because of the bright red staining caused by oxidized iron coatings on the grains. Widespread salt lakes formed, and salt deposits accumulated.

As the period continued, more of the land was flooded, and limestones were formed. At the very end

(Below) The world as it may have looked in Early Triassic times.

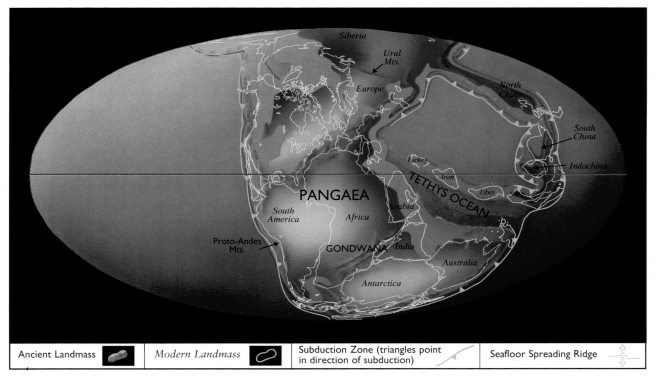

Ancient Landmass	*Modern Landmass*	Subduction Zone (triangles point in direction of subduction)	Seafloor Spreading Ridge

---

46

of the Triassic a great subsidence took place in the area called Tethys (later to become the Alps and the Himalayas) and also around the edge of what later became the Pacific Ocean.

(Below) The Petrified Forest National Park in Arizona is of Triassic age.

Triassic life

The great extinction of many species of life that occurred at the end of the Permian Period meant that the seas no longer contained the brachiopods, crinoids, corals, and trilobites. A few new species of coral (hexacorals) began to develop, but were very rare. In the place of the brachiopods clam-type bivalves evolved steadily, while ammonites were able to diversify from what had been a single group, although they were almost made extinct again at the end of the Triassic Period.

Amphibians also increased during the Triassic Period, only to be almost wiped out again at the end. Reptiles were more fortunate, increasing in numbers and not faring as badly during the extinctions at the end of the period.

Mammal-like reptiles also existed at this time, although most were very small.

Triassic rocks and minerals

During the Triassic Period some areas had a hot, humid climate, and the swamps were buried to form coals. Coal deposits thick enough for mining occur in Spitzbergen, North China, Australia, South Africa, the eastern United States, and Brazil. Oil also developed in the limited seas, producing some of the reserves of the North Sea and Alaskan North Slope. On the other hand, salt and gypsum formed over wide areas. There are no metal-bearing ores formed in Triassic times because there was no widespread mountain building or volcanic activity.

The Jurassic Period

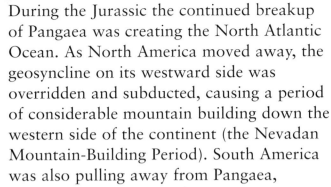

During the Jurassic the continued breakup of Pangaea was creating the North Atlantic Ocean. As North America moved away, the geosyncline on its westward side was overridden and subducted, causing a period of considerable mountain building down the western side of the continent (the Nevadan Mountain-Building Period). South America was also pulling away from Pangaea, creating a range of mountains down its western flank. Meanwhile, geosyclines also formed on the trailing edges of these continents. But there was no overriding, and so the geosynclines that began forming, for example, off the eastern coast of North America in Jurassic times are still forming today.

As continents pulled away from one another, new seafloor material welled up in the form of basalt lava flows. At the same time, the moving continents consumed large areas of old ocean floor. For this reason no ocean floor anywhere in the world is older than the Jurassic Period.

(Below) The world as it may have looked in Early Jurassic times.

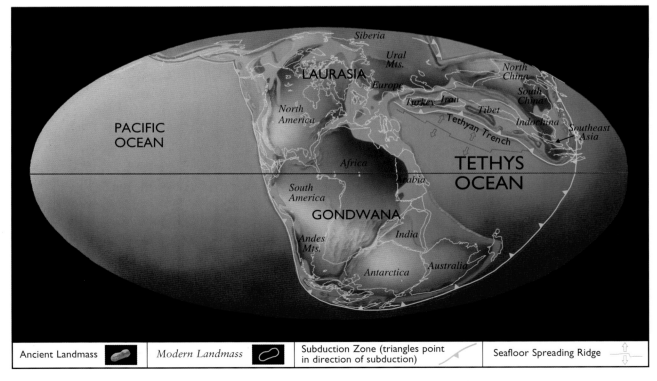

Ancient Landmass	Modern Landmass	Subduction Zone (triangles point in direction of subduction)	Seafloor Spreading Ridge

Widespread subsidence occurred during the Jurassic Period, allowing the sea to flood in and cover large areas of all continents with shallow waters. In these seas a multitude of life flourished, providing the basis for widespread limestones to form. More limestones accumulated on most coastal edges, with especially thick deposits along the shore of the Tethys Ocean. They would later be thrust up to make many of the mountains of southern Europe and Asia, including the Jura and the Alps.

(Below) Dinosaur National Monument rocks are of Jurassic age.

Jurassic life

With another extinction at the end of the Triassic, Jurassic life had to rebuild from a low base. The climate appears to have been warmer and wetter than in Triassic times, and this allowed large forests to spread out over the continents. Today they are represented by coal deposits in China, Siberia, Iran, Mexico, and British Columbia.

The dinosaurs were reptiles that survived the Triassic extinction and took great advantage of the warmer and wetter conditions, diversifying greatly. In this way the largest land animals of all time (the brachiosaurs), sea reptiles such as ichthyosaurs, and the first plesiosaurs all appeared. In the air pterosaurs became common. More significant for later periods, the first bird, *Archaeopteryx*, is found in the Late Jurassic.

Sharks and bony fish survived through the extinction and were the main Jurassic fishes. Ammonites, nearly wiped out at the end of the Triassic Period, again grew in number and range of species. Their rapid evolution, and the fact that they were able to travel worldwide, makes them, rather than any land-based reptiles, the fossils used for dividing the Jurassic system. Tightly coiled ammonites characterize the Lower Jurassic, with more loosely coiled species with greater ornamentation found in the Middle and Upper Jurassic. Belemnites were also common in Jurassic seas, together with echinoids and bivalves. The new hexacorals were now able to build widespread reefs.

Jurassic rocks and minerals

Coal, oil, and natural gas are all important. Most of the Middle East oil and gas reserves and those of the North Sea were formed during the Jurassic Period. Mountain building on the western side of North America produced conditions for vulcanism, and minerals were concentrated there, particularly gold deposits in California that were later to be the source of a gold rush.

In the less turbulent conditions of Eurasia subsidence and flooding not only produced large areas of limestone, but also concentrated iron reserves to create huge sheets of iron ore that were later used as the basis of the Industrial Revolution in Europe.

The Cretaceous Period

Many rocks of Cretaceous age are still found at the surface on many continents because it was a period when there was much new rock formation, and also because it is relatively recent, and so there has not been time for the rocks to erode away.

The continued breakup of Pangaea was still having important effects on the world's climates as well as on mountain building.

Chalk was formed during the Cretaceous Period mainly from coccoliths (the calcite secretions of some kinds of algae).

At the beginning of the Cretaceous the continental land of the earth was divided into two large pieces—Gondwana and Laurasia, although narrow seas had already opened between North America and the rest of Laurasia, and between South America and the rest of Gondwana. One of the important events was the closing of the Tethys Ocean, while the spread of the continents was making the Pacific Ocean smaller, and the Atlantic Ocean larger.

(Below) The world as it may have looked in the Late Cretaceous.

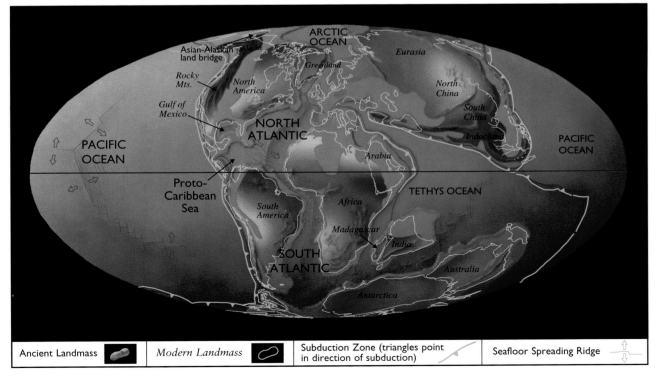

| Ancient Landmass | | Modern Landmass | | Subduction Zone (triangles point in direction of subduction) | Seafloor Spreading Ridge |

All around the leading edges of the continent continued subsidence of the coast (in geosynclines) allowed material eroded from the land to build up into great thicknesses of sediment. The continents also subsided, allowing the seas to flood in. Sea levels were higher in the Cretaceous Period than at any other time in history, being about 200 meters higher than at present. The reason for this is thought to be that the midoceanic ridges were growing all over the earth at this time in what were mainly very narrow oceans.

The formation of huge underwater mountains displaced the ocean waters onto the land. (Later, as the oceans widened, the sea level fell again.) In some places these seas were clear, and extensive limestones and chalks were formed. In other, deeper places oil-rich shales were formed. For example, as the sea flooded across the middle of what is now North America, oil-rich shale formed in the north of the sea, in what is now Montana and Wyoming. Farther south, where the sea was shallower, chalk formed from Kansas to Texas.

Cretaceous rocks and minerals

The warm, shallow seas and marginal swamps of the Cretaceous provided opportunities for the formation of coal. Thus the Cretaceous is an important period for coal deposits. They are mainly mined in western North America and northeastern Asia. Similarly, it was an ideal environment for the formation of oil, and half of all the world's oil deposits are of Cetaceous age, with three-quarters of them being formed in the Middle East, and most of the rest forming in the area centered on the Gulf of Mexico, Venezuela, the Rocky Mountains, and Australia.

Subduction and mountain building during this period created the right conditions for the concentration of many important metals, mostly wherever volcanic activity and magma chambers occurred. Gold, silver, copper, lead, and zinc are mined in western North America and the Andes, and diamonds from the kimberlite deposits of South Africa.

Cretaceous life

The Cretaceous Period saw the decline of many older species of tree and the rise of the flowering plants (angiosperms). Conifers, which once had broad distribution, gave way to the new seed-bearing evergreen and deciduous trees, and occupied only the colder, drier, or less fertile areas. Insects began to be abundant, but the mammals were still relatively few in number.

Flying reptiles were still common at the start of the Cretaceous Period, but birds were beginning to become more dominant. Plesiosaurs and other large marine reptiles continued to thrive, and on land dinosaurs continued to be the dominant life form. *Tyrannosaurus*, the largest of the land-based carnivores, continued to roam until the Late Cretaceous.

Some new dinosaurs evolved, while others became extinct. The last of them were the duck-billed dinosaurs.

Ammonites continued to evolve in the Cretaceous, although they, like the dinosaurs, all died out at the end of the Cretaceous. Meanwhile, small algae continued to flourish, their calciferous shells providing the coccoliths that make up the chalk rocks.

The leading edges of the continents were continually involved in subduction, just as they had been in the Jurassic. This caused extensive mountain building throughout the Cretaceous Period down the entire western margin of the Americas. As the continental margin crumpled, the effects spread farther and farther east, until by the end of the period they were causing renewed uplift as far east as the Rocky Mountains. As these mountains rose, erosion speeded up, and sediments began to fill in eastward over the shales and chalks. This would eventually be the region now known as the High Plains, just east of the Rockies.

Volcanic activity was very common, and in India huge fissure eruptions allowed basalt to flood over the land on a scale never witnessed before, creating basaltic sheets (the Deccan Traps) that cover 500,000 square kilometers.

The same pattern of inland seas was also occurring in Europe. At this time chalk was deposited from England eastward into the continent. Today's London and Paris Basins are rimmed and underlain with chalk from this time. Inland seas also affected central Australia.

(Above) Cretaceous sandstone at Mesa Verde National Park, Colorado.

(Below) The Cretaceous Period was the time of widespread chalk formation, of which the White Cliffs of Dover, England, and nearby areas are the most famous landmark.

The Cenozoic Era

The Cenozoic is the most recent of the three major eras of geological time. Because events have happened so recently, many of the rocks formed have not had time to be properly consolidated, so that many of them are soft and easily eroded. However, it has also been a time of considerable plate movement, and it is in the Cenozoic Era that many of the world's major mountain systems have been formed.

The Cenozoic Era has two major divisions, very unequal in length: the Tertiary Period, which is everything from the end of the Cretaceous Period to the Ice Age, and the Quaternary Period, which is the short period from the beginning of the Ice Age to the present.

Much of the modern shape of the continents and their landscapes has been influenced by events in the Cenozoic. The Americas moved west, while the Tethys Ocean closed, and the Indian Ocean opened.

Major events took place on all continents. In North America the Appalachian Mountains, which had been reduced almost to plains, were lifted up, the Rocky Mountains and the Sierra Nevada were also lifted higher, and the crust stretched out to make the Basin and Range country of the southwestern United States. The Pacific Plate began to move northward, and the North American plate stopped riding over it. The result was the formation of the San Andreas Fault through California. New volcanic mountains formed in the Pacific Northwest, where subduction continued. They formed the Cascade Mountains as well as the mountains of Mexico and Central America. Broad uplifting of the land in the southwestern United States caused the Colorado River to begin to cut downward, forming the

Cenozoic life

The mass extinctions that took place at the end of the Cretaceous Period removed dinosaurs from their dominance on earth, and the Cenozoic Era is sometimes called the Age of Mammals, although it has also been a time when all kinds of other living things—such as plants, insects, and birds—have also flourished.

The modern forms of life developed in the Cenozoic. Birds, fish, and mammals, which had existed for 100 million years on earth, suddenly became common. Some of the mammals were much larger than those today. Uintatheres and titanotheres were 9 meters long and stood 6 meters at the shoulder. There were also 6-meter-high pigs, 5-meter-tall sloths, woolly mammoths, and saber-tooth cats.

In the seas ammonites became extinct, and bivalves and new forms of coral became abundant.

Primates developed in the last part of the Cenozoic Era, the rapidly improving technology of humans leading to the extinction of many species by the end of the Pleistocene Epoch.

Grand Canyon. Basalts flooded over the area now drained by the Columbia and Snake rivers, and the Hawaiian Islands were formed.

The European Alps, the Atlas Mountains of Africa, and the Asian Himalayas were created by India and Africa colliding with the southern edge of Eurasia. The Red Sea formed as Arabia pulled away from Africa. The East African Rift Valley was also formed at this time. The crust of the western side of Europe was stretched, resulting in swarms of dikes and sills in Scotland, Ireland, and northern England, including the sill on which Hadrian's Wall was built in Roman times.

At the beginning of the Cenozoic Era the climate was much warmer than it is today, but toward the end the earth became enveloped in a major ice age.

(Below) The Himalayas were raised to great heights as India (left) collided with Asia (right) during the Tertiary Mountain-Building Period.

The Tertiary Period

The Tertiary Period was the time during which mammals became the dominant life form on earth. At this time the climate was warmer than it is today. Most of the continents were nearly in their modern positions, but there were some land connections that do not exist today. Especially important was the land bridge between North America and Asia, which meant that animals could cross between continents more easily than they can today. No similar land bridge existed in the southern continents, and as a result, life in the southern continents evolved to be more different than in the north.

There have been no major flooding periods in the Tertiary Period, and so the rocks of this period are only found around the edges of the present continents. They fill, for example, the central regions of the London, Paris, and Mississippi basins.

Although the closing of the Tethys Ocean had been going on for many millions of years, the final rise of the Alps and Himalayas did not take place

(Below) The world as it may have looked in Miocene times.

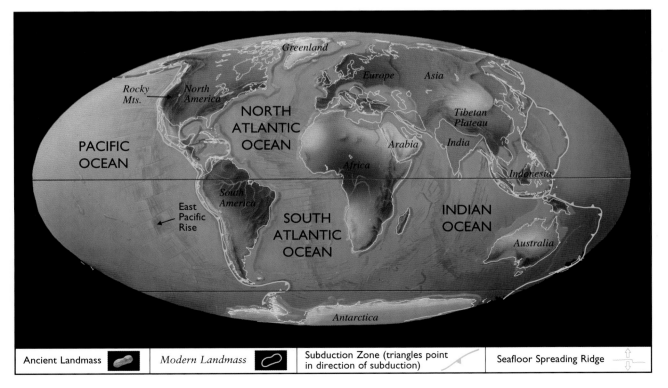

| Ancient Landmass | | *Modern Landmass* | | Subduction Zone (triangles point in direction of subduction) | Seafloor Spreading Ridge |

until the middle of the Tertiary Period. The rising continents changed the world's climates, resulting in the typical formation of deserts in the places we find them today.

The Alps and the Himalayas rose at different times. The Alps grew earlier as a result of the northward movement of Africa. The Himalayas emerged in the Mid-Tertiary as India moved north, but the high Tibetan Plateau did not begin to rise until even later.

Volcanic events were widespread in North America, including the basalts of the Columbia-Snake region of the Northwest. In the oceans the Mid-Atlantic Ridge and the East-Pacific Rise continued to grow and become the world's major (though unseen) mountain ranges. In India the Deccan Traps region was flooded with basalts. Huge explosive volcanoes formed in northwestern Scotland, Ireland, Greenland, and northeastern North America. This was caused by the final rifting apart of the most northerly part of North America from Europe. More explosive volcanoes occurred as the East African Rift Valley began to form. By the Tertiary Period the **PACIFIC RING OF FIRE** was well developed, with volcanoes

Tertiary life

At the end of the Mesozoic a large number of species became extinct. Plants survived relatively unscathed, but both land and ocean animals suffered considerably. Of the plants, grasses have been the main group to expand and diversify.

The most important change was that there was a great reduction in those animals, for example, algae, that produced calcium carbonate and were responsible for the Cretaceous chalk deposits. Many reptiles went extinct, and mammals became the largest life form on earth.

The recovery of the species that survived was slow, but new forms of corals eventually grew up. Bivalves and gastropods became important in the marine environment, along with a wide variety of fishes. Sharks survived, and whales (whose ancestors had evolved on land) adapted to the sea and became the largest mammal, while new kinds of birds, such as penguins, developed.

(*Left*) The Cascade Mountains were formed in the northwestern United States. The volcanoes that now rise from them are Quaternary in age. This is Mount Rainier in Washington State.

marking the subduction zone of the Pacific Ocean Plate under the enclosing continents. The ripple effect of this subduction caused not just the reuplift of the Rocky Mountains and the Colorado Plateau, but widespread volcanic activity. Deep downwarping of basins in the western United States allowed the accumulation of thousands of meters of Tertiary sediments.

At the beginning of the Tertiary Period subduction occurred under the whole of the west coast of North America, but gradually, subduction stopped in the south, so that volcanic activity then ceased, continuing only in the Cascades to the north. Instead, major rift systems developed, of which the San Andreas Fault is the largest.

Hot spots under the crust have produced many chains of volcanoes, such as the Hawaiian Islands, the Galapagos, and the Society Islands.

In New Zealand (which by this time lay across a major plate boundary) the Kaikoura Mountain-Building Period began and resulted in the formation of the Southern Alps.

Tertiary rocks

Great contrasts have occurred in the rocks forming during Tertiary times. In places close to the newly rising mountains huge sheets of coarse sediments were laid down. Elsewhere beds of poorly consolidated sands and clays were deposited. Volcanic activity was common in places where the continents were pulling apart, such as in Scotland and in the Basin and Range area of the United States. HOT SPOTS created huge outpourings of lava to cover areas such as the Deccan Traps in India. In warm regions close to the equator corals began to build reefs.

(Below) The world as it may have looked in Pleistocene times.

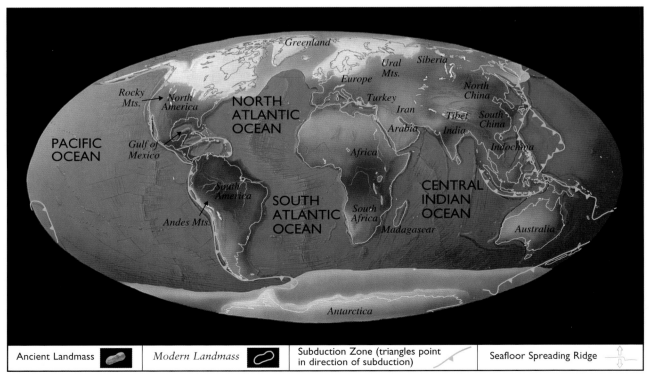

| Ancient Landmass | | *Modern Landmass* | | Subduction Zone (triangles point in direction of subduction) | | Seafloor Spreading Ridge | |

The Quaternary Period

The Quaternary Period is the most recent part of geological history, being less than two million years long. The main interval in the Quaternary is called the Pleistocene Epoch, and it lasted until about 11,000 years ago. The main event during this time has been the Ice Age. The postglacial period is called the Holocene Epoch.

The Pleistocene saw the gradual buildup of colder conditions and the expansion of ice sheets over all the northern continents, both in the mountains and on the plains. The ocean water locked up on land as ice during this time caused sea levels to fall by about 80 meters. Many of the continental shelves were therefore exposed, and rivers and glaciers cut valleys into this new land.

Wherever the ice sheets and glaciers flowed, rock was scoured away. Much of this material – a mixture

(Below) The Ice Age has been the main feature of Quaternary times. Ice still covers the northern and southernmost continents. This is Greenland.

of rock flour and boulders and known as boulder-clay or glacial till—was finally deposited as broad, hummocky spreads when the ice melted. This material masks much of the rock in the northern hemisphere, roughly from a line that goes through New York and London.

As the ice melted, the seas rose, flooding into the lower parts of the valleys. Modern estuaries, rias, and fjords all owe their origins to this rise in sea level. In areas where the thickest ice was present, the removal of its burden led to a rebound of the land. As it rose, previous shorelines became dry land, and the floors of river valleys were also lifted up. Old shorelines are seen as wave-cut platforms around many coasts, and old valley floors as benches in valleys called river terraces.

Quaternary life

The fluctuations in climate and the spread of glaciation had enormous effects on living things. As conditions became colder, the distribution of plants changed and moved equatorward. The result was that all the bands of natural vegetation around the world became narrower. Animals moved with their sources of food, leaving vast areas near the north as frozen wastes.

As the climate warmed different areas of land, the plants began to advance poleward once more, their seeds successfully growing farther and farther north each season. This process is still continuing today, and species continue to migrate northward since the retreat of the last phase of glaciation.

(Below) Geological changes are still occurring at the same rate as they have in the past. We see this mainly through active volcanoes (this picture shows El Misti in Peru) and through earthquakes.

Glossary

aa lava: a type of lava with a broken, bouldery surface.

abrasion: the rubbing away (erosion) of a rock by the physical scraping of particles carried by water, wind, or ice.

acidic rock: a type of igneous rock that consists predominantly of light-colored minerals and more than two-thirds silica (e.g., granite).

active volcano: a volcano that has observable signs of activity, for example, periodic plumes of steam.

adit: a horizontal tunnel drilled into rock.

aftershock: an earthquake that follows the main shock. Major earthquakes are followed by a number of aftershocks that decrease in frequency with time.

agglomerate: a rock made from the compacted particles thrown out by a volcano (e.g., tuff).

alkaline rock: a type of igneous rock containing less than half silica and normally dominated by dark-colored minerals (e.g., gabbro).

amygdule: a vesicle in a volcanic rock filled with such secondary minerals such as calcite, quartz, or zeolite.

andesite: an igneous volcanic rock. Slightly more acidic than basalt.

anticline: an arching fold of rock layers in which the rocks slope down from the crest. *See also* syncline.

Appalachian Mountain (Orogenic) Belt: an old mountain range that extends for more than 3,000 km along the eastern margin of North America from Alabama in the southern United States to Newfoundland, Canada, in the north. There were three Appalachian orogenies: Taconic (about 460 million years ago) in the Ordovician; Acadian (390 to 370 million years ago) in the Devonian; and Alleghenian (300 to 250 million years ago) in the Late Carboniferous to Permian. These mountain belts can be traced as the Caledonian and Hercynian orogenic belts in Europe.

Archean Eon: *see* eon.

arenaceous: a rock composed largely of sand grains.

argillaceous: a rock composed largely of clay.

arkose: a coarse sandstone formed by the disintegration of a granite.

ash, volcanic: fine powdery material thrown out of a volcano.

asthenosphere: the weak part of the upper mantle below the lithosphere, in which slow convection is thought to take place.

augite: a dark green-colored silicate mineral containing calcium, sodium, iron, aluminum, and magnesium.

axis of symmetry: a line or plane around which one part of a crystal is a mirror image of another part.

basalt: basic fine-grained igneous volcanic rock; lava often contains vesicles.

basic rock: an igneous rock (e.g., gabbro) with silica content less than two-thirds and containing a high percentage of dark-colored minerals.

basin: a large, circular, or oval sunken region on the earth's surface created by downward folding. A river basin, or watershed, is the area drained by a river and its tributaries.

batholith: a very large body of plutonic rock that was intruded deep into the earth's crust and is now exposed by erosion.

bauxite: a surface material that contains a high percentage of aluminum silicate. The principal ore of aluminum.

bed: a layer of sediment. It may involve many phases of deposition, each marked by a bedding plane.

bedding plane: an ancient surface on which sediment built up. Sedimentary rocks often split along bedding planes.

biotite: a black-colored form of mica.

body wave: a seismic wave that can travel through the interior of the earth. P waves and S waves are body waves.

boss: an upward extension of a batholith. A boss may once have been a magma chamber.

botryoidal: the shape of a mineral that resembles a bunch of grapes, e.g., hematite whose crystals are often arranged in massive clumps, giving a surface covered with spherical bulges.

butte: a small mesa.

calcareous: composed mainly of calcium carbonate.

calcite: a mineral composed of calcium carbonate.

caldera: the collapsed cone of a volcano. It sometimes contains a crater lake.

Caledonian Mountain-Building Period, Caledonian Orogeny: a major mountain-building period in the Lower Paleozoic Era that reached its climax at the end of the Silurian Period (430 to 395 million years ago). An early phase affected only North America and made part of the Appalachian Mountain Belt.

Cambrian, Cambrian Period: the first period of geological time in the Paleozoic Era, beginning 570 million years ago and ending 500 million years ago.

carbonate minerals: minerals formed with carbonate ions (e.g., calcite).

Carboniferous, Carboniferous Period: a period of geological time between about 345 and 280 million years ago. It is often divided into the Early Carboniferous Epoch (345 to 320 million years ago) and the Late Carboniferous Epoch (320 to 280 million years ago). The Late Carboniferous is characterized by large coal-forming swamps. In North America the Carboniferous is usually divided into the Mississippian (= Lower Carboniferous) and Pennsylvanian (= Upper Carboniferous) periods.

cast, fossil: the natural filling of a mold by sediment or minerals that were left when a fossil dissolved after being enclosed by rock.

Cenozoic, Cenozoic Era: the most recent era of geological time, beginning 65 million years ago and continuing to the present.

central vent volcano: *see* stratovolcano

chemical compound: a substance made from the chemical combination of two or more elements.

chemical rock: a rock produced by chemical precipitation (e.g., halite).

chemical weathering: the decay of a rock through the chemical action of water containing dissolved acidic gases.

cinder cone: a volcanic cone made entirely of cinders. Cinder cones have very steep sides.

class: the level of biological classification below a phylum.

clast: an individual grain of a rock.

clastic rock: a sedimentary rock that is made up of fragments of preexisting rocks, carried by gravity, water, or wind (e.g., conglomerate, sandstone).

cleavage: the tendency of some minerals to break along one or more smooth surfaces.

coal: the carbon-rich, solid mineral derived from fossilized plant remains. Found in sedimentary rocks. Types of coal include bituminous, brown, lignite, and anthracite. A fossil fuel.

complex volcano: a volcano that has had an eruptive history that produces two or more vents.

composite volcano: *see* stratovolcano.

concordant coast: a coast where the geological structure is parallel to the coastline. *See also* discordant coastline.

conduction (of heat): the transfer of heat between touching objects.

conglomerate: a coarse-grained sedimentary rock with grains larger than 2 mm.

contact metamorphism: metamorphism that occurs due to direct contact with a molten magma. *See also* regional metamorphism.

continental drift: the theory suggested by Alfred Wegener that earth's continents were originally one land mass that split up to form the arrangement of continents we see today.

continental shelf: the ocean floor from the coastal shore of continents to the continental slope.

continental shield: the ancient and stable core of a tectonic plate. Also called a shield.

convection: the slow overturning of a liquid or gas that is heated from below.

cordillera: a long mountain belt consisting of many mountain ranges.

core: the innermost part of the earth. The earth's core is very dense, rich in iron, partly molten, and the source of the earth's magnetic field. The inner core is solid and has a radius of about 1,300 kilometers. The outer core is fluid and is about 2,100 kilometers thick. S waves cannot travel through the outer core.

cracking: the breaking up of a hydrocarbon compound into simpler constituents by means of heat.

crater lake: a lake found inside a caldera.

craton: *see* shield.

Cretaceous, Cretaceous Period: the third period of the Mesozoic Era. It lasted between about 135 and 65 million years ago. It was a time of chalk formation and when many dinosaurs lived.

cross-bedding: a pattern of deposits in a sedimentary rock in which many thin layers lie at an angle to the bedding planes, showing that the sediment was deposited by a moving fluid. Wind-deposited cross-beds are often bigger than water-deposited beds.

crust: the outermost layer of the earth, typically 5 km thick under the oceans and 50 to 100 km thick under continents. It makes up less than 1 percent of the earth's volume.

crustal plate: *see* tectonic plate.

crystal: a mineral that has a regular geometric shape and is bounded by smooth, flat faces.

crystal system: a group of crystals with the same arrangement of axes.

crystalline: a mineral that has solidified but been unable to produce well-formed crystals. Quartz and halite are commonly found as crystalline masses.

crystallization: the formation of crystals.

cubic: a crystal system in which crystals have 3 axes all at right angles to one another and of equal length.

cuesta: a ridge in the landscape formed by a resistant band of dipping rock. A cuesta has a steep scarp slope and a more gentle dip slope.

current bedding: a pattern of deposits in a sedimentary rock in which many thin layers lie at an angle to the bedding planes, showing that the sediment was deposited by a current of water.

cyclothem: a repeating sequence of rocks found in coal strata.

delta: a triangle of deposition produced where a river enters a sea or lake.

deposit, deposition: the process of laying down material that has been transported in suspension or solution by water, ice, or wind. A deposit is the material laid down by deposition (e.g., salt deposits).

destructive plate boundary: a line where plates collide and one plate is subducted into the mantle.

Devonian, Devonian Period: the fourth period of geological time in the Paleozoic Era from 395 to 345 million years ago.

dike: a wall-like sheet of igneous rock that cuts across the layers of the surrounding rocks.

dike swarm: a collection of hundreds or thousands of parallel dikes.

diorite: an igneous plutonic rock between gabbro and granite; the plutonic equivalent of andesite.

dip: the angle that a bedding plane or fault makes with the horizontal.

dip slope: the more gently sloping part of a cuesta whose surface often parallels the dip of the strata.

discontinuity: a gap in deposition, perhaps caused by the area being lifted above the sea so that erosion, rather than deposition, occurred for a time.

discordant coast: a coast where the rock structure is at an angle to the line of the coast. *See also* concordant coastline.

displacement: the distance that one piece of rock is pushed relative to another.

dissolve: to break down a substance into a solution without causing a reaction.

distillation: the boiling off of volatile materials, leaving a residue.

dolomite: a mineral composed of calcium magnesium carbonate.

dome: a circular uplifted region of rocks taking the shape of a dome and found in some areas of folded rocks. Rising plugs of salt will also dome up the rocks above them. They sometimes make oil traps.

dormant volcano: a volcano that shows no signs of activity but that has been active in the recent past.

drift: a tunnel drilled in rock and designed to provide a sloping route for carrying out ore or coal by means of a conveyor belt.

earthquake: shaking of the earth's surface caused by a sudden movement of rock within the earth.

element: a fundamental chemical building block. A substance that cannot be separated into simpler substances by any chemical means. Oxygen and sulfur are examples of elements.

eon: the largest division of geological time. An eon is subdivided into eras. Precambrian time is divided into the Archean (earlier than 2.5 billion years ago) and Proterozoic eons (more recent than 2.5 billion years ago). The Phanerozoic Eon includes the Cambrian Period to the present.

epicenter: the point on the earth's surface directly above the focus (hypocenter) of an earthquake.

epoch: a subdivision of a geological period in the geological time scale (e.g., Pleistocene Epoch).

era: a subdivision of a geological eon in the geological time scale (e.g., Cenozoic Era). An era is subdivided into periods.

erode, erosion: the twin processes of breaking down a rock (called weathering) and then removing the debris (called transporting).

escarpment: the crest of a ridge made of dipping rocks.

essential mineral: the dominant mineral constituents of a rock used to classify it.

evaporite: a mineral or rock formed as the result of evaporation of salt-laden water, such as a lagoon or salt lake.

exoskeleton: another word for shell. Applies to invertebrates.

extinct volcano: a volcano that has shown no signs of activity in historic times.

extrusive rock, extrusion: an igneous volcanic rock that has solidified on the surface of the earth.

facet: the cleaved face of a mineral. Used in describing jewelry.

facies: physical, chemical, or biological variations in a sedimentary bed of the same geological age (e.g., sandy facies, limestone facies).

family: a part of the classification of living things above a genus.

fault: a deep fracture or zone of fractures in rocks along which there has been displacement of one side relative to the other. It represents a weak point in the crust and upper mantle.

fault scarp: a long, straight, steep slope in the landscape that has been produced by faulting.

feldspar: the most common silicate mineral. It consists of two forms: plagioclase and orthoclase.

ferromagnesian mineral: dark-colored minerals such as augite and hornblende that contain relatively high proportions of iron and magnesium and low proportions of silica.

fissure: a substantial crack in a rock.

fjord: a glaciated valley in a mountainous area coastal area that has been partly flooded by the sea.

focal depth: the depth of an earthquake focus below the surface.

focus: the origin of an earthquake, directly below the epicenter.

fold: arched or curved rock strata.

fold axis: line following the highest arching in an anticline or the lowest arching in a syncline.

fold belt: a part of a mountain system containing folded sedimentary rocks.

foliation: a texture of a rock (usually schist) that resembles the pages in a book.

formation: a word used to describe a collection of related rock layers, or beds. A number of related beds make a member; a collection of related members makes up a formation. Formations are often given location names, e.g., Toroweap Formation, whose members are a collection of dominantly limestone beds.

fossil: any evidence of past life, including remains, traces, and imprints.

fossil fuel: any fuel that was formed in the geological past from the remains of living organisms. The main fossil fuels are coal and petroleum (oil and natural gas).

fraction: one of the components of crude oil that can be separated from others by heating and then cooling the vapor.

fracture: a substantial break across a rock.

fracture zone: a region in which fractures are common. Fracture zones are particularly common in folded rock and near faults.

frost shattering: the process of breaking pieces of rock through the action of freezing and melting of rainwater

gabbro: alkaline igneous plutonic rock, typically showing dark-colored crystals; plutonic equivalent of basalt.

gallery: a horizontal access tunnel in a mine.

gangue: the unwanted mineral matter found in association with a metal.

gem: a mineral, usually in crystal form, that is regarded as having particular beauty and value.

genus: (*pl.* genera) the biological classification for a group of closely related species.

geode: a hollow lump of rock (nodule) that often contains crystals.

geological column: a columnar diagram showing the divisions of geological time (eons, eras, periods, and epochs).

geological eon: *see* eon.

geological epoch: *see* epoch.

geological era: *see* era.

geological period: a subdivision of a geological era (e.g., Carboniferous Period). A period is subdivided into epochs.

geological system: a term for an accumulation of strata that occurs during a geological period (e.g., the Ordovician System is the rocks deposited during the Ordovician Period). Systems are divided into series.

geological time: the history of the earth revealed by its rocks.

geological time scale: the division of geological time into eons, era, periods, and epochs.

geosyncline: a large, slowly subsiding region marginal to a continent where huge amounts of sediment accumulate. The rocks in a geosyncline eventually are lifted to form mountain belts.

gneiss: a metamorphic rock showing large grains.

graben: a fallen block of the earth's crust forming a long trough separated on all sides by faults. Associated with rift valleys.

grain: a particle of a rock or mineral.

granite: an acidic, igneous, plutonic rock containing free quartz, typically light in color; plutonic equivalent of rhyolite.

grit: grains larger than sand but smaller than stones.

groundmass: *see* matrix.

group: a word used to describe a collection of related rock layers, or beds. A number of related beds make a member; a collection of related members makes up a formation; a collection of related formations makes a group.

gypsum: a mineral made of calcium sulfate.

halide minerals: a group of minerals (e.g., halite) that contain a halogen element (elements similar to chlorine) bonded with another element. Many are evaporite minerals.

halite: a mineral made of sodium chloride.

Hawaiian-type eruption: a name for a volcanic eruption that mainly consists of lava fountains.

hexagonal: a crystal system in which crystals have 3 axes all at 120 degrees to one another and of equal length.

hogback: a cuesta where the scarp and dip slopes are about the same angle.

hornblende: a dark-green silicate mineral of the amphibole group containing sodium, potassium, calcium, magnesium, iron, and aluminum.

horst: a raised block of the earth's crust separated on all sides by faults. Associated with rift valleys.

hot spot: a place where a fixed mantle magma plume reaches the surface.

hydraulic action: the erosive action of water pressure on rocks.

hydrothermal: a change brought about in a rock or mineral due to the action of superheated mineral-rich fluids, usually water.

hypocenter: the calculated location of the focus of an earthquake.

ice wedging: *see* frost shattering

Icelandic-type eruption: a name given to a fissure type of eruption.

igneous rock: rock formed by the solidification of magma. Igneous rocks include volcanic and plutonic rocks.

impermeable: a rock that will not allow a liquid to pass through it.

imprint: a cast left by a former life form.

impurities: small amounts of elements or compounds in an otherwise homogeneous mineral.

index fossil: a fossil used as a marker for a particular part of geological time.

intrusive rock, intrusion: rocks that have formed from cooling magma below the surface. When inserted among other rocks, intruded rocks are called an intrusion.

invertebrate: an animal with an external skeleton.

ion: a charged particle.

island arc: a pattern of volcanic islands that follow the shape of an arc when seen from above.

isostacy: the principle that a body can float in a more dense fluid. The same as buoyancy, but used for continents.

joint: a significant crack between blocks of rock, normally used in the context of patterns of cracks.

Jurassic, Jurassic Period: the second geological period in the Mesozoic Era, lasting between 190 and 135 million years ago.

kingdom: the broadest division in the biological classification of living things.

laccolith: a lens-shaped body of intrusive igneous rock with a dome-shaped upper surface and a flat bottom surface.

landform: a recognizable shape of part of the landscape, for example, a cuesta.

landslide: the rapid movement of a slab of soil down a steep hillslope.

lateral fault: *see* thrust fault.

laterite: a surface deposit containing a high proportion of iron.

lava: molten rock material extruded onto the surface of the earth.

lava bomb: *see* volcanic bomb.

law of superposition: the principle that younger rock is deposited on older.

limestone: a carbonate sedimentary rock composed of more than half calcium carbonate.

lithosphere: that part of the crust and upper mantle that is brittle and makes up the tectonic plates.

lode: a mining term for a rock containing many rich ore-bearing minerals. Similar to vein.

Love wave, L wave: a major type of surface earthquake wave that shakes the ground surface at right angles to the direction the wave is traveling in. It is named after A.E.H. Love, the English mathematician who discovered it.

luster: the way in which a mineral reflects light. Used as a test when identifying minerals.

magma: the molten material that comes from the mantle and that cools to form igneous rocks.

magma chamber: a large cavity melted in the earth's crust and filled with magma. Many magma chambers are plumes of magma that have melted their way from the mantle to the upper part of the crust. When a magma chamber is no longer supplied with molten magma, the magma solidifies to form a granite batholith.

mantle: the layer of the earth between the crust and the core. It is approximately 2,900 kilometers thick and is the largest of the earth's major layers.

marginal accretion: the growth of mountain belts on the edges of a shield.

mass extinction: a time when the majority of species on the planet were killed off.

matrix: the rock or sediment in which a fossil is embedded; the fine-grained rock in which larger particles are embedded, for example, in a conglomerate.

mechanical weathering: the disintegration of a rock by frost shattering/ice wedging.

mesa: a large detached piece of a tableland.

Mesozoic, Mesozoic Era: the geological era between the Paleozoic and the Cenozoic eras. It lasted between 225 and 65 million years ago.

metamorphic aureole: the region of contact metamorphic rock that surrounds a batholith.

metamorphic rock: any rock (e.g., schist, gneiss) that was formed from a preexisting rock through heat and pressure.

meteorite: a substantial chunk of rock in space.

micas: a group of soft, sheetlike silicate minerals (e.g., biotite, muscovite).

midocean ridge: a long mountain chain on the ocean floor where basalt periodically erupts, forming new oceanic crust.

mineral: a naturally occurring inorganic substance of definite chemical composition (e.g., calcite, calcium carbonate).
  More generally, any resource extracted from the ground by mining (includes, metal ores, coal, oil, gas, rocks, etc.).

mineral environment: the place where a mineral or a group of associated minerals forms. Mineral environments include igneous, sedimentary, and metamorphic rocks.

mineralization: the formation of minerals within a rock.

Modified Mercalli Scale: a scale for measuring the impact of an earthquake. It is composed of 12 increasing levels of intensity that range from imperceptible, designated by Roman numeral I, to catastrophic destruction, designated by XII.

Mohorovicic discontinuity: the boundary surface that separates the earth's crust from the underlying mantle. Named for Andrija Mohorovicic, a Croatian seismologist.

Mohs' Scale of Hardness: a relative scale developed to put minerals into an order. The hardest is 10 (diamond), and the softest is 1 (talc).

mold: an impression in a rock of the outside of an organism.

monoclinic: a crystal system in which crystals have 2 axes all at right angles to one another, and each axis is of unequal length.

mountain belt: a region where there are many ranges of mountains. The term is often applied to a wide belt of mountains produced during mountain building.

mountain building: the creation of mountains as a result of the collision of tectonic plates. Long belts or chains of mountains can form along the edge of a continent during this process. Mountain building is also called orogeny.

mountain-building period: a period during which a geosyncline is compressed into fold mountains by the collision of two tectonic plates. Also known as orogenesis.

mudstone: a fine-grained, massive rock formed by the compaction of mud.

nappe: a piece of a fold that has become detached from its roots during intensive mountain building.

native metal: a metal that occurs uncombined with any other element.

natural gas: *see* petroleum.

normal fault: a fault in which one block has slipped down the face of another. It is the most common kind of fault and results from tension.

nueé ardente: another word for pyroclastic flow.

ocean trench: a deep, steep-sided trough in the ocean floor caused by the subduction of oceanic crust beneath either other oceanic crust or continental crust.

olivine: the name of a group of magnesium iron silicate minerals that have an olive color.

order: a level of biological classification between class and family.

Ordovician, Ordovician Period: the second period of geological time within the Paleozoic Era. It lasted from 500 to 430 million years ago.

ore: a rock containing enough useful metal or fuel to be worth mining.

ore mineral: a mineral that occurs in sufficient quantity to be mined for its metal. The compound must also be easy to process.

organic rocks: rocks formed by living things, for example, coal.

orthoclase: the form of feldspar that is often pink in color and that contains potassium as important ions.

orogenic belt: a mountain belt.

orogeny: a period of mountain building. Orogenesis is the process of mountain building and the creation of orogenic belts.

orthorhombic: a crystal system in which crystals have 3 axes all at right angles to one another but of unequal length.

outcrop: the exposure of a rock at the surface of the earth.

overburden: the unwanted layer(s) of rock above an ore or coal body.

oxide minerals: a group of minerals in which oxygen is a major constituent. A compound in which oxygen is bonded to another element or group.

Pacific Ring of Fire: the ring of volcanoes and volcanic activity that circles the Pacific Ocean. Created by the collision of the Pacific Plate with its neighboring plates.

pahoehoe lava: the name for a form of lava that has a smooth surface.

Paleozoic, Paleozoic Era: a major interval of geological time. The Paleozoic is the oldest era in which fossil life is commonly found. It lasted from 570 to 225 million years ago.

paleomagnetism: the natural magnetic traces that reveal the intensity and direction of the earth's magnetic field in the geological past.

pegmatite: an igneous rock (e.g., a dike) of extremely coarse crystals.

Pelean-type eruption: a violent explosion dominated by pyroclastic flows.

period: *see* geological period.

permeable rock: a rock that will allow a fluid to pass through it.

Permian, Permian Period: the last period of the Paleozoic Era, lasting from 280 to 225 million years ago.

petrified: when the tissues of a dead plant or animal have been replaced by minerals, such as silica, they are said to be petrified (e.g., petrified wood).

petrified forest: a large number of fossil trees. Most petrified trees are replaced by silica.

petroleum: the carbon-rich, and mostly liquid, mixture produced by the burial and partial alteration of animal and plant remains. Petroleum is found in many sedimentary rocks. The liquid part of petroleum is called oil. The gaseous part of petroleum is known as natural gas. Petroleum is an important fossil fuel.

petroleum field: a region from which petroleum can be recovered.

Phanerozoic Eon: the most recent eon, beginning at the Cambrian Period, some 570 million years ago, and extending up to the present.

phenocryst: an especially large crystal (in a porphyritic rock) embedded in smaller mineral grains.

phylum: (*pl.* phyla) biological classification for one of the major divisions of animal life and second in complexity to kingdom. The plant kingdom is not divided into phyla but into divisions.

placer deposit: a sediment containing heavy metal grains (e.g., gold) that have weathered out of the bedrock and concentrated on a stream bed or along a coast.

plagioclase: the form of feldspar that is often white or gray and that contains sodium and calcium as important ions.

planetismals: small embryo planets.

plate: *see* plate tectonics, tectonic plate.

plateau: an extensive area of raised flat land. The clifflike edges of a plateau may, when eroded, leave isolated features such as mesas and buttes. *See also* tableland.

plate tectonics: the theory that the earth's crust and upper mantle (the lithosphere) are broken into a number of more or less rigid, but constantly moving, slabs or plates.

Plinian-type eruption: an explosive eruption that sends a column of ash high into the air.

plug: *see* volcanic plug.

plunging fold: a fold whose axis dips, or plunges, into the ground.

plutonic rock: an igneous rock that has solidified at great depth and contains large crystals due to the slowness of cooling (e.g., granite, gabbro).

porphyry, porphyritic rock: an igneous rock in which larger crystals (phenocrysts) are enclosed in a fine-grained matrix.

Precambrian, Precambrian time: the whole of earth history before the Cambrian Period. Also called Precambrian Era and Precambrian Eon.

precipitate: a substance that has settled out of a liquid as a result of a chemical reaction between two chemicals in the liquid.

Primary Era: an older name for the Paleozoic Era.

prismatic: a word used to describe a mineral that has formed with one axis very much longer than the others.

Proterozoic Eon: *see* eon.

P wave, primary wave, primary seismic wave: P waves are the fastest body waves. The waves carry energy in the same line as the direction of the wave. P waves can travel through all layers of the earth and are generally felt as a thump. *See also* S wave.

pyrite: iron sulfide. It is common in sedimentary rocks that were poor in oxygen and sometimes forms fossil casts.

pyroclastic flow: solid material ejected from a volcano, combined with searingly hot gases, which together behave as a high-density fluid. Pyroclastic flows can do immense damage, as was the case with Mount Saint Helens.

pyroclastic material: any solid material ejected from a volcano.

Quaternary, Quaternary Period: the second period in the Cenozoic Era, beginning about 1.6 million years ago and continuing to the present day.

radiation: the transfer of energy between objects that are not in contact.

radioactive dating: the dating of a material by the use of its radioactive elements. The rate of decay of any element changes in a predictable way, allowing a precise date to be given since the material was formed.

rank: a name used to describe the grade of coal in terms of its possible heat output. The higher the rank, the more the heat output.

Rayleigh wave: a type of surface wave having an elliptical motion similar to the waves caused when a stone is dropped into a pond. It is the slowest, but often the largest and most destructive, of the wave types caused by an earthquake. It is usually felt as a rolling or rocking motion and, in the case of major earthquakes, can be seen as they approach. Named after Lord Rayleigh, the English physicist who predicted its existence.

regional metamorphism: metamorphism resulting from both heat and pressure. It is usually connected with mountain building and occurs over a large area. *See also* contact metamorphism.

reniform: a kidney-shaped mineral habit (e.g., hematite).

reservoir rock: a permeable rock in which petroleum accumulates.

reversed fault: a fault where one slab of the earth's crust rides up over another. Reversed faults are only common during plate collision.

rhyolite: acid, igneous, volcanic rock, typically light in color; volcanic equivalent of granite.

ria: the name for a partly flooded coastal river valley in an area where the landscape is hilly.

Richter Scale: the system used to measure the strength of an earthquake. Developed by Charles Richter, an American, in 1935.

rift, rift valley: long troughs on continents and midocean ridges that are bounded by normal faults.

rifting: the process of crustal stretching that causes blocks of crust to subside, creating rift valleys.

rock: a naturally occurring solid material containing one or more minerals.

rock cycle: the continuous sequence of events that causes mountains to be formed then eroded before being formed again.

rupture: the place over which an earthquake causes rocks to move against one another.

salt dome: a balloon-shaped mass of salt produced by salt being forced upward under pressure.

sandstone: a sedimentary rock composed of cemented sand-sized grains 0.06–2mm in diameter.

scarp slope: the steep slope of a cuesta.

schist: a metamorphic rock characterized by a shiny surface of mica crystals all oriented in the same direction.

scoria: the rough, often foamlike rock that forms on the surface of some lavas.

seamount: a volcano that rises from the seabed.

Secondary Era: an older term for a geological era. Now replaced by Mesozoic Era.

sediment: any solid material that has settled out of suspension in a liquid.

sedimentary rock: a layered clastic rock formed through the deposition of pieces of mineral, rock, animal, or vegetable matter.

segregation: the separation of minerals.

seismic gap: a part of an active fault where there have been no earthquakes in recent times.

seismic wave: a wave generated by an earthquake.

series: the rock layers that correspond to an epoch of time.

shadow zone: the region of the earth that experiences no shocks after an earthquake.

shaft: a vertical tunnel that provides access or ventilation to a mine.

shale: a fine-grained sedimentary rock made of clay minerals with particle sizes smaller than 2 microns.

shield: the ancient and stable core of a tectonic plate. Also called a continental shield.

shield volcano: a volcano with a broad, low-angled cone made entirely from lava.

silica, silicate: silica is silicon dioxide. It is a very common mineral, occurring as quartz, chalcedony, etc. A silicate is any mineral that contains silica.

sill: a tabular, sheetlike body of intrusive igneous rock that has been injected between layers of sedimentary or metamorphic rock.

Silurian, Silurian Period: the name of the third geological period of the Paleozoic Era. It began about 430 and ended about 395 million years ago.

skarn: a mineral deposit formed by the chemical reaction of hot acidic fluids and carbonate rocks.

slag: waste rock material that becomes separated from the metal during smelting.

slate: a low-grade metamorphic rock produced by pressure, in which the clay minerals have arranged themselves parallel to one another.

slaty cleavage: a characteristic pattern found in slates in which the parallel arrangement of clay minerals causes the rock to fracture (cleave) in sheets.

species: a population of animals or plants capable of interbreeding.

spreading boundary: a line where two plates are being pulled away from each other. New crust is formed as molten rock is forced upward into the gap.

stock: a vertical protrusion of a batholith that pushes up closer to the surface.

stratigraphy: the study of the earth's rocks in the context of their history and conditions of formation.

stratovolcano: a tall volcanic mountain made of alternating layers, or strata, of ash and lava.

stratum: (*pl.* strata) a layer of sedimentary rock.

streak: the color of the powder of a mineral produced by rubbing the mineral against a piece of unglazed, white porcelain. Used as a test when identifying minerals.

striation: minute parallel grooves on crystal faces.

strike, direction of: the direction of a bedding plane or fault at right angles to the dip.

Strombolian-type eruption: a kind of volcanic eruption that is explosive enough

to send out some volcanic bombs.

subduction: the process of one tectonic plate descending beneath another.

subduction zone: the part of the earth's surface along which one tectonic plate descends into the mantle. It is often shaped in the form of an number of arcs.

sulfides: a group of important ore minerals (e.g., pyrite, galena, and sphalerite) in which sulfur combines with one or more metals.

surface wave: any one of a number of waves such as Love waves or Rayleigh waves that shake the ground surface just after an earthquake. *See also* Love waves and Rayleigh waves.

suture: the junction of 2 or more parts of a skeleton; in cephalopods the junction of a septum with the inner surface of the shell wall. It is very distinctive in ammonoids and used to identify them.

S wave, shear or secondary seismic wave: this kind of wave carries energy through the earth like a rope being shaken. S waves cannot travel through the outer core of the earth because they cannot pass through fluids. *See also* P wave.

syncline: a downfold of rock layers in which the rocks slope up from the bottom of the fold. *See also* anticline.

system: see geological system.

tableland: another word for a plateau. *See* plateau.

tectonic plate: one of the great slabs, or plates, of the lithosphere (the earth's crust and part of the earth's upper mantle) that cover the whole of the earth's surface. The earth's plates are separated by zones of volcanic and earthquake activity.

Tertiary, Tertiary Period: the first period of the Cenozoic Era. It began 665 and ended about 1.6 million years ago.

thrust fault: *see* reversed fault.

transcurrent fault: *see* lateral fault.

transform fault: *see* lateral fault.

translucent: a description of a mineral that allows light to penetrate but not pass through.

transparent: a description of a mineral that allows light to pass right through.

trellis drainage pattern: a river drainage system where the trunk river and its tributaries tend to meet at right angles.

trench: *see* ocean trench.

Triassic, Triassic Period: the first period of the Mesozoic era. It lasted from about 225 to 190 million years ago.

triclinic: a crystal system in which crystals have 3 axes, none at right angles or of equal length to one another.

tsunami: a very large wave produced by an underwater earthquake.

tuff: a rock made from volcanic ash.

unconformity: any interruption in the depositional sequence of sedimentary rocks.

valve: in bivalves and brachiopods, one of the separate parts of the shell.

vein: a sheetlike body of mineral matter (e.g., quartz) that cuts across a rock. Veins are often important sources of valuable minerals. Miners call such important veins lodes.

vent: the vertical pipe that allows the passage of magma through the center of a volcano.

vertebrate: an animal with an internal skeleton.

vesicle: a small cavity in a volcanic rock originally created by an air bubble trapped in the molten lava.

viscous, viscosity: sticky, stickiness.

volatile: substances that tend to evaporate or boil off of a liquid.

volcanic: anything from or of a volcano. Volcanic rocks are igneous rocks that cool as they are released at the earth's surface— including those formed underwater; typically have small crystals due to the rapid cooling, e.g., basalt, andesite, and rhyolite.

volcanic bomb: a large piece of magma thrown out of a crater during an eruption, which solidifies as it travels through cool air.

volcanic eruption: an ejection of ash or lava from a volcano.

volcanic glass: lava that has solidified very quickly and has not had time to develop any crystals. Obsidian is a volcanic glass.

volcanic plug: the solidified core of an extinct volcano.

Vulcanian-type eruption: an explosive form of eruption without a tall ash column or pyroclastic flow.

water gap: a gap cut by a superimposed river, which is still occupied by the river.

weather, weathered, weathering: the process of weathering is the mechanical action of ice and the chemical action of rainwater on rock, breaking it down into small pieces that can then be carried away. *See also* chemical weathering and mechanical weathering.

wind gap: a gap cut by a superimposed river, which is no longer occupied by the river.

Set Index

USING THE SET INDEX

This index covers all eight volumes in the *Earth Science* set:

Volume
number   Title

- 1: **Minerals**
- 2: **Rocks**
- 3: **Fossils**
- 4: **Earthquakes and volcanoes**
- 5: **Plate tectonics**
- 6: **Landforms**
- 7: **Geological time**
- 8: **The earth's resources**

An example entry:

Index entries are
listed alphabetically.

plagioclase feldspar **1:** *51*; **2:** 10 *see also*
     feldspars

Volume numbers are in bold and are followed by page references. Articles on a subject are shown by italic page numbers.

In the example above, "plagioclase feldspar" appears in Volume 1: Minerals on page 51 as a full article and in Volume 2: Rocks on page 10. Many terms also are covered in the GLOSSARY on pages 60–65.

The *see also* refers to another entry where there will be additional relevant information.

A

aa lava **2:** 24; **4:** 37, 44
abrasion **6:** 41
Aconcagua, Mount (Argentina) **5:** 38
Adirondacks, Adirondack Mountains (New York) **7:** 27
adit **8:** 39, 46
African Plate **5:** 11, 50
African Shield **5:** 54, 55, 56
aftershocks **4:** 14, 15
agate **1:** *50*
Agathla Peak (Arizona) **6:** 55
agglomerate **2:** 18
Agricola **1:** 30
Alaska **5:** 35
     1964 earthquake **4:** 9, 10, 12, *22–25*
     oil fields **8:** 37
Alethopteris **3:** 49
Aleutian Islands (Alaska) **5:** 34, 35
algae **3:** 48, 51
Alleghenian Mountain-Building Period **7:** 41
almandine garnet **1:** 52
Alps **5:** 44, 45–51; **7:** 54, 56

aluminum **1:** 7; **8:** 8, 11, 33
     ores **8:** *12*
amber, fossils in **3:** 10, 38, 59
amethyst **1:** 5, *48*, 49
ammonites **3:** 4, 15, 16, 24, *26–28*, 56, 58, 59
     recognizing fossils **3:** 27
ammonoids **3:** *26–28*, 54, 55, 56, 57, 58
Ampato (Peru) **5:** 37
amphibians **3:** 42, 46, 54, 55, 56
amphiboles **1:** *56*; **2:** 12
amygdules **2:** 16
Anchorage (Alaska) **4:** 22, 23, 25
andalusite **1:** *52*
Andes Mountains, Andes (South America) **5:** 11, *36–38*
andesite **2:** 16, 18, 21, *23*; **4:** 38, 39
Angara Shield **5:** 57
anhydrite **1:** *46*
Annularia **3:** 49
Antarctic Plate **5:** 11
Antarctica **5:** 45
anthracite **8:** 28, 29
anticlines **6:** 37, 38, 39; **8:** 32
antimony **8:** 10
apatite **1:** 23, *47*
apophylite **1:** 24
Appalachian Mountain Belt **5:** 45
Appalachian Mountains, Appalachians (United States) **5:** 44, 45; **6:** 36, 37, 38, 39, 40; **7:** 29, 30, 41, 53
     coal fields **8:** 35
aquamarine **1:** 25, *55*
Arabian Plate **5:** 11
aragonite **1:** 28, *44*
Archaeopteryx **3:** 58
Archean Eon **7:** 25
Arches National Park (Utah) **6:** 20
arches, coastal **6:** *41–46*
archosaurs **3:** 57
Arduino, Giovanni **7:** 15
arenaceous rocks **2:** 42
argillaceous sandstone **2:** 43
arkose sandstone **2:** 42, 44
arsenic **1:** 31; **8:** 10, 45, 57
arthropods (Arthropoda) **3:** *38–41*, 51, 53, 54
Articulatean **3:** 48
Arun River (Nepal) **6:** 10
ash, volcanic **2:** 16, 18, 19; **4:** 35, 36; **6:** 24, 55
asthenosphere **5:** 6, 8, 9
Atlantic Ocean **5:** 23, 25; **7:** 50, 55, 57
     oil fields **8:** 37
Atlantis, research ship **5:** 25
Atlas Mountains (Morocco) **7:** 54
atmosphere, formation **7:** *24–27*
augen gneiss **2:** 59
augite **1:** 7, 14, 16, *56*; **2:** 12
Australian Shield **5:** 54, 56, 57
Ayers Rock (Australia) **5:** 57
azurite **1:** 20, *44*

B

Baltic Shield **5:** 57; **7:** 25
Baltica **7:** 30, 32, 34, 36, 39
banded gneiss **2:** 59
barite **1:** 27, *46*
basalt **1:** 12; **2:** 14, 15, 17, 20, 21, *23–25*; **4:** 37, 38, 39, 42, 44, 46, 47, 48, 49; **5:** 25; **6:** 53, 54; **8:** 8

basalt columns **2:** 15, 25; **6:** 18, 26
Basin and Range **4:** 27, 28; **5:** 42; **6:** 49
basins **6:** 40
batholiths **1:** 11; **6:** 19, 52, 53, 59
     ore minerals **8:** 24
Bauer, Georg **1:** 30
bauxite **1:** *39*; **8:** *12*, 33, 41
bedding, bedding planes **2:** 32, 36, 37; **6:** 20, 32
belemnites (Belemnoids) **3:** 4, *24–25*, 57, 58, 59
     recognizing fossils **3:** 24
Bendigo (Australia) **7:** 35; **8:** 43
Bennetitalean **3:** 48
beryl **1:** 27, *55*
Berzelius, Jons **1:** 30
Big Bend National Park (Texas) **5:** 13; **6:** 58
biotite **1:** *58*; **2:** 11 *see also* mica
birds **3:** 42, 57, 58
bismuth **1:** 31; **8:** 10
bituminous coal **8:** 28, 29, 36*(map)*
bivalves (Bivalvia) **3:** *29–31*, 52, 56
     recognizing fossils **3:** 30
bloodstone **1:** 50
Blue Ridge Mountains (United States) **5:** 45
body waves **4:** *8 see also* P and S waves
bony fish **3:** 44, 54
Bora Bora (French Polynesia) **4:** 43
borate minerals **1:** 45
borax **1:** *45*; **8:** 27
bornite **1:** *36*; **8:** *15*
boss **6:** 52, 53
brachiopods (Brachiopoda) **3:** *21–23*, 52, 53, 54, 55, 56
     recognizing **3:** 21
Brazilian Shield **5:** 56
breccia **2:** *40–41*
Bristol Lake (California) **8:** 27
brittle stars **3:** 56
Broken Hill (Australia) **8:** *25*
brown coal *see* lignite
Bryce Canyon (Utah) **6:** 21; **7:** 5, 10, 13
Bushveld Complex (South Africa) **8:** 21
buttes **6:** 26, 27, 34

C

Cabot Fault (Canada) **6:** 50
cadmium **8:** 11
calcareous skeleton **3:** 6
calcite **1:** 18, 19, 23, *44*; **2:** 13
calderas **4:** *58–59*; **5:** 23; **6:** 52, 57
Caledonian Mountain-Building Period **7:** 35
Caledonian Mountains **5:** 45; **7:** 30, 35, 37, 38
California Valley **5:** 39
Calymene **3:** 41, 54
calyx **3:** 34, 35
Cambrian, Cambrian Period **3:** 50*(chart)*; **7:** 18*(name)*, 21*(chart)*, *32(map)–33*
     fossils **3:** 52
     life **7:** 33
     rocks **7:** 33
Canadian Rockies **5:** 40, 41 *see also* Rockies
Canadian Shield **5:** 56; **7:** 25, 27
Canyon del Colca (Peru) **5:** 37
canyons **6:** 26, 27
cap rocks **6:** *26–27*, 30, 32, 53; **8:** 32
carbon **1:** 31
carbonate minerals **1:** *44–45*; **8:** 11, 53